Analytical Chemistry Experiment

分析化学实验

（英汉双语版）

主　编　王卫平

副主编　吴　靓　刘卫东　翁雪香　钱兆生

科学出版社

北　京

内 容 简 介

本书共 5 章，第 1 章主要介绍分析化学实验基本知识；第 2 章主要介绍定量分析基本操作；第 3 章是定量分析实验(含 32 个实验项目)；第 4 章是分光光度法及常用分离方法实验(含 7 个实验项目)；第 5 章是综合设计实验(含 18 个实验项目)。本书编排了一些与所学知识高度相关且紧密结合生活实际的综合设计实验，以培养学生分析问题、解决问题的能力。

本书可供高等学校化学、化工、生物、食品、环境等相关专业师生使用，也可供从事化学实验工作的人员参考。

图书在版编目(CIP)数据

分析化学实验=Analytical Chemistry Experiment：英、汉/王卫平主编. —北京：科学出版社，2019.8

ISBN 978-7-03-061182-6

Ⅰ. ①分…　Ⅱ. ①王…　Ⅲ. ①分析化学–化学实验–高等学校–教材–英、汉　Ⅳ. ①O652.1

中国版本图书馆 CIP 数据核字(2019)第 089827 号

责任编辑：丁　里/责任校对：何艳萍

责任印制：张　伟/封面设计：迷底书装

科学出版社出版

北京东黄城根北街 16 号

邮政编码：100717

http://www.sciencep.com

北京厚诚则铭印刷科技有限公司 印刷

科学出版社发行　各地新华书店经销

*

2019 年 8 月第 一 版　开本：787×1092　1/16

2023 年11月第四次印刷　印张：16

字数：420 000

定价：59.00 元

(如有印装质量问题，我社负责调换)

《分析化学实验》(英汉双语版)

编写委员会

主　编　王卫平

副主编　吴　靓　刘卫东　翁雪香　钱兆生

编　委(按姓名汉语拼音排序)

冯九菊　刘卫东　钱兆生　王卫平

翁雪香　吴　靓　吴兰菊　吴小华

袁军华　袁培新　张　露　仲淑贤

前　　言

化学是一门实践性很强的学科。分析化学实验是分析化学课程的重要组成部分。学生通过分析化学实验课程的学习，可以加深对分析化学基本理论的理解，正确和熟练地掌握分析化学实验的基本方法和基础操作技能，培养实事求是、严谨的科学态度。

为应对教育及人才国际化的趋势，浙江师范大学于 2014、2015 年分别批准生物技术和应用化学两个专业为国际化专业。分析化学实验是这两个专业的学位必修课，被列为全英文授课教学的课程，但却苦于缺少一本适用的英文教材，本书的编写应运而生。本书参照国内较为成熟的分析化学实验内容，结合浙江师范大学近年来国际化专业建设过程中“分析化学实验”全英文课程授课的经验编写而成。全书分 5 章，第 1 章主要介绍分析化学实验基本知识；第 2 章主要介绍定量分析基本操作；第 3 章是定量分析实验(含 32 个实验项目)；第 4 章是分光光度法及常用分离方法实验(含 7 个实验项目)；第 5 章是综合设计实验(含 18 个实验项目)。编者在编写过程中查阅了大量的文献资料，增加了一些与实验教学内容高度相关且紧密结合生活实际的综合设计实验，有助于学生掌握基本操作技能，同时激发其学习兴趣，培养其解决实际问题的能力，逐步养成严谨、求实、创新的科学态度。

本书是浙江师范大学分析化学学科全体教师多年来教学实践的结晶，凝结了集体创作的智慧。参加编译的成员有：王卫平(第 1、2 章及实验 1、7、17～19、28、40～54、57 编译及全文校对统稿)；吴靓(实验 2～6 编译及全文校对统稿)；翁雪香(实验 25、27、29、33、34 编译及实验 23～35 校对)；刘卫东(实验 11～14、55、56 编译)；钱兆生(实验 20～22、36～39 编译校对)；袁军华(实验 8、15、16、23、24、26 编译)；张露(实验 30、31、35 编译)；袁培新(实验 9、10、32 编译)；吴兰菊(第 1、2 章及实验 1～10 校对)。此外，在前期素材收集及编写过程中，吴小华、冯九菊、仲淑贤、王雪等提出了许多宝贵建议并做了大量工作；后期排版工作主要由分析化学专业研究生张莹同学负责完成。在此向他们表示衷心的感谢！在本书编写过程中，得到了浙江师范大学化学与生命科学学院的大力支持，同时得到了许多同行、专家的指导和帮助，在此表示衷心的感谢！

由于编者水平所限，书中难免有一些疏漏，恳请广大教师和读者批评指正。

编　者
2018 年 12 月

前 言

目　录

Chapter 1 Basic Knowledge of Analytical Chemistry Experiments

1.1 Basic Requirements for Analytical Chemistry Experiments

As compulsory basic course for students majoring in chemistry and related fields, analytical chemistry experiments plays a key role in studying analytical chemistry. Through the study of this course, students can deepen their understanding of the basic concepts and basic principles of analytical chemistry; master the basic operations and skills of analytical chemistry experiments; cultivate their rigorous and meticulous work style and the scientific attitude of seeking truth from facts. To learn this course well, you need to do the following:

(1) Prepare carefully before the experiment. Understand the experimental principle, familiarize with the experimental content, methods, procedures and precautions based on experimental teaching materials and theoretical study. Write a preview report, and list the data recording form.

(2) During the experiment, the operation is strictly regulated, and the experimental phenomena should be carefully observed and recorded in a timely and truthful manner. Students should have a notebook for data recording. The data should not be recorded on a single sheet of paper or small pieces of paper. The text recording should be neat and clean. The data records should be expressed in a regular manner in tabular form. Don't make fakes and make up data.

(3) The experiment table and the laboratory should be kept tidy and quiet during the experiment. After the experiment is completed, the platform should be cleaned in time, and the instruments and reagents should be placed in an orderly manner.

(4) Finish an experiment report in time. After the experiment is over, the original report should be carefully sorted, analyzed, and the experiment report should be completed independently. The experimental report generally includes the following items: experiment name, the date, objectives, principles, equipment and reagents, contents and procedures, data recording and processing, questions and discussions.

1.2 Basic Knowledge of the Laboratory

1.2.1 Laboratory Safety Knowledge

In analytical chemistry experiments, corrosive, flammable and explosive chemical reagents, easily broken glass instruments, some precision instruments, water, electricity and gas are often used. To ensure the personal safety of the experimenter and the normal conduct of the experiment, the following laboratory safety rules must be strictly obeyed.

(1) Eating and smoking are strictly prohibited in the laboratory.

(2) Wash hands after the experiment. Shut down water, electricity and gas immediately after use, and check again before leaving the laboratory.

(3) Operate various instruments after the teacher explains the demonstration or reads the operating procedures.

(4) Do not use wet hands to open the switch to avoid electric shock when using electrical equipment.

(5) When using concentrated acid, alkali or other strong corrosive agents, do not splash it on skin and clothes. When using concentrated HNO_3, HCl, H_2SO_4, $HClO_4$ or ammonia, it should be carried out in a fume hood. Make sure to cool concentrated ammonia and HCl with tap water before opening the cap in summer.

(6) When using flammable organic solvents (such as ether, acetone, ethanol, benzene, etc.), be sure to keep away from flames and heat sources. Immediately cover the cap after it is used and store in a cool place. Low-boiling organic solvents are prohibited from heating directly on open flames and can only be heated in a water bath.

(7) Hot, concentrated $HClO_4$ is easily exploded in the presence of organic compound. If the sample is organic, it should be heated with concentrated nitric acid to react with the organic matter, then add $HClO_4$ after the organic matter is destroyed. The smoke generated by evaporation of $HClO_4$ is easily condensed in the fume hood. If $HClO_4$ is used frequently, the fume hood should be washed regularly with water to prevent the condensation of $HClO_4$ from dust and organic matter, causing combustion or explosion.

(8) Special care should be taken when using highly toxic substances such as mercury salts, arsenic compounds and cyanide. Pay special attention to the fact that cyanide should not be exposed to acid, otherwise it will produce highly toxic HCN! The cyanide-containing waste liquid can be adjusted to pH > 10 by adding NaOH first, and then add excessive bleaching powder for oxidative decomposition of CN^-; or add an excessive ferrous sulfate solution under alkaline conditions to convert CN^- to ferrocyanide. The compound is then treated as a waste liquid. It is strictly forbidden to pour cyanide-containing waste liquid directly into the sewer or waste tank.

Hydrogen sulfide gas is toxic. Operations involving hydrogen sulfide gas must be carried out in a fume hood.

(9) If scald occurs, wipe the yellow bitter acid solution or burn ointment on the scalded skin. Severe cases should be sent to hospital immediately. In the event of a fire in the laboratory, targeted fire suppression should be carried out according to the cause of the fire. When organic solvents such as gasoline and ether are on fire, use sand to extinguish. At this time, never use water. When the wire or electric appliance is on fire, do not use water or CO_2 fire extinguisher. Instead, first cut off the power supply, use CCl_4 fire extinguisher to extinguish the fire, and decide whether to report to the fire department.

(10) The laboratory should keep clean and tidy. Do not throw the brush or rag in the water tank. It is forbidden to throw solid objects, glass fragments, etc. into the water tank to avoid blockage of the sewer. Such materials, as well as waste paper should be placed in the waste bin or in the

place where the laboratory is required to store. Waste acid and alkali should be carefully poured into the waste liquid tank. Do not pour into the water tank to avoid corrosion of the water pipe.

1.2.2 Water for Analytical Chemistry Experiments

1. Specifications of Water for Analytical Chemistry Experiments

Pure water is commonly used as solvent and detergent in analytical chemistry experiments. Different types of pure water should be selected according to the experimental requirements. The national standard of water for analytical laboratory (GB/T 6682—2008), providing technical specifications, preparation and test methods, has been established in China. Level and main indicators of pure water used in a laboratory are listed in Table 1.1.

Table 1.1 Level and main indicators of pure water used in a laboratory

Name	Class Ⅰ	Class Ⅱ	Class Ⅲ
pH range (25℃)	—	—	5.0~7.5
Conductivity (25℃)/(mS/m)	≤ 0.01	≤ 0.10	≤ 0.50
Oxidizable material (O)/(mg/L)	—	≤ 0.08	≤ 0.40
Absorbance (254 nm, optical path of 1 cm)	≤ 0.001	≤ 0.01	—
Evaporites[(105 ± 2)℃]/(mg/L)	—	≤ 1.0	≤ 2.0
Soluble silicon (SiO_2)/(mg/L)	≤ 0.01	≤ 0.02	—

In practice, some experiments have special requirements for pure water, which can be selected according to the relevant items to be tested, such as oxygen, iron and ammonia content, etc.

2. Preparation of Pure Water

The water used to dissolve, dilute and prepare solution must be purified in analytical laboratory. Different analytical experiment requires different quality of the pure water. Thus, different purification methods should be used according to the experimental requirements. Pure water used in a laboratory contains distilled water, deionized water, conductivity water, double distilled water, CO_2-free distilled water, ammonia-free distilled water, etc. The preparation methods are as follows.

1) Distilled Water

Heat tap water in an evaporator and then condense to obtain distilled water. Impurity ions is generally not volatile, so distilled water contains much less impurities than tap water. The quality of distilled water can reach Class Ⅲ, but it contains a small amount of metal ions, CO_2 and other impurities.

2) Deionized Water

Deionized water is obtained from tap water or ordinary distilled water through the column of ion exchange resin. The purity of deionized water is higher than that of distilled water. The quality of deionized water can reach Class Ⅱ or Ⅰ. But it may contain non-electrolyte, colloidal substances and trace amounts of organic matter.

3) Conductivity Water

Distilled water and a small amount of potassium permanganate are added into the first distiller. Organic matter is removed by distillation to obtain the double distilled water. Then the double distilled water is transferred to the second distiller with a small amount of barium sulfate and potassium bisulfate for distillation. 10 mL of both head and end distillation are discarded. Conductivity water is obtained by collecting mid distillation.

4) Double Quartz Subboiling Distilled Water

Double distillation is necessary to obtain pure water. Add appropriate reagents to inhibit the volatilization of certain impurities. Generally, the quality of double distilled water can reach Class Ⅱ.

5) Special Water

Ammonia-free distilled water: add 25 mL of 5% NaOH solution to per liter of distilled water and boil it for 1 h. Check ammonium ion or add 2 mL of concentrated H_2SO_4 to per liter of distilled water and redistill.

CO_2-free distilled water: boil distilled water until it is 3/4 or 4/5 of original volume, isolate from air and cool. CO_2-free distilled water should be stored in a bottle with an absorption tube containing soda lime. And the pH should be kept at 7.

Chlorine-free distilled water: boil distilled water in a hard glass distiller and distill to collect mid distillation.

3. Test of Water Purity

Conductivity is the main indicator of water quality. Appropriate instrument should be used to measure the conductivity of pure water. The concentration of certain impurities can be determined according to specifically experimental requirements. General test items are as follows.

1) Resistivity

At 25℃, the resistivity of pure water and ultrapure water are $(1.0\sim10)\times10^6$ and $10\times10^6\ \Omega\cdot cm$, respectively.

2) pH

The pH of pure water for analytical experiments is 6～7.

3) Ca^{2+} and Mg^{2+}

Drop ammonia-ammonium chloride buffer solution to appropriate amount of water to be tested, adjust the pH to about 10 and add 1 drop of Eriochrome black T indicator, no red color appears.

According to specific requirements, testing items for pure water also include chloride, iron, copper ions, etc.

1.2.3 Washing of Glassware

1. Cleaning Method

Washing of apparatus is basic and technical for chemical workers. The washing degree of apparatus used for quantitative analysis is directly related to the precision and accuracy of analytical results.

The washing methods should be selected according to the experimental requirements, the properties of dirt and the degree of contamination. The methods commonly used are as follows.

1) First Used Glassware

There are always some alkaline substances on the wall of glassware newly bought. They should be washed using diluted detergent solution, soapy water or detergent powder and then be washed with tap water. After washing, they should be immersed in 1%～2% HCl solution for overnight or not less than 4 h, then be washed with tap water, distilled water 2～3 times and dried in an oven at 80～100℃.

2) General Glassware

For general glassware such as test tubes, beakers, Erlenmeyer flasks, graduated cylinder, etc., first wash them with tap water until there is no dirt. Then brush the outside and inside of vessels carefully by a suitable size brush using diluted detergent solution. After rinsing with tap water inside and outside, rinse with distilled water 2～3 times, dry or place upside down at the clean place. Any clean glassware should not contain water droplets on the wall, otherwise the glassware should be washed again as described above. If the inner wall is found to have smudges that are difficult to remove, it should be re-washed.

3) Graduated Vessel

For graduated vessel such as pipette, burette, volumetric flasks etc., they should be immersed in cool water immediately after use. Rinse off the attached reagents, proteins and other substances after the experiment is finished and dry them. Then soak them in the acidic washing solution for 4～6 h (or overnight), rinse them thoroughly with tap water, and rinse with distilled water for 2～4 times. Finally, dry in the air for use.

4) Other Containers

Containers containing various toxic drugs, especially highly toxic drugs and radioisotopes, must be specially treated to ensure that no residual poisons exist before they can be cleaned.

2. Commonly Used Detergent

1) Chromic Acid Lotion

The chromic acid lotion is a solution containing saturated $K_2Cr_2O_7$ in concentrated sulfuric acid. It has strong oxidizing properties and is suitable for washing inorganic substances, oil stains and some organic substances. Preparations: 5 g of crude $K_2Cr_2O_7$ is finely ground and dissolved in 10 mL of hot water. After continuous stirring, 100 mL of industrial grade concentrated H_2SO_4 is added slowly. The solution is dark red. After cooling, it is transferred to a glass bottle and set aside. The chromic acid lotion can be used repeatedly. When the solution is green, indicating that the lotion has failed and must be reconstituted. The chromic acid lotion is very corrosive, and the hexavalent chromium is harmful to the human body. Pay attention to safety when using it.

2) NaOH-$KMnO_4$ Solution

NaOH-$KMnO_4$ Solution is used for washing oil stains and certain organic substances. Preparations: dissolve 4 g of $KMnO_4$ in water and add 100 mL of 100 g/L NaOH solution.

3) Synthetic Detergent

Synthetic detergent is used for washing oil stains and certain organic substances. They are mainly washing powder, detergent, etc.

4) Hydrochloric Acid-ethanol Solution

Hydrochloric acid-ethanol solution is used for washing cuvettes, volumetric flasks and pipettes contaminated by colored materials. The chemical pure hydrochloric acid and ethanol are mixed in a volume ratio of 1∶2.

5) Acidic Oxalic Acid and Hydroxylamine Hydrochloride Solution

It is suitable for washing oxidizing substances, such as containers contaminated with $KMnO_4$, MnO_2, Fe^{3+}, etc. Preparations: 10 g of oxalic acid or 1 g of hydroxylamine hydrochloride is dissolved in 100 mL of 1∶1 hydrochloric acid solution.

1.2.4 Chemical Reagent

1. Grade of Chemical Reagents

The purity of chemical reagent has great influence on the experimental results. Different experiments have different demands on the purity of chemical reagent. Therefore, the classification criteria of reagents should be understood. Grade of chemical reagents is shown in Table 1.2.

Table 1.2 Grade of chemical reagents

Grade	Name	Sign	Label color	Scope of application
first grade	guaranteed reagent	GR	green	precise analytical experiments
second grade	analytical reagent	AR	red	general analytical experiments
third grade	chemical reagent	CP or P	blue	general chemical experiments
biochemical reagent	biological reagent	BR or CR	yellow	biochemical experiments

Chemical reagents (CP or P) with few impurities are used for general chemistry experiments. Analytical reagents (AR) with much fewer impurities are used for assay determination. Moreover, there are some reagents for special use. For example, the concentration of impurities in "spectral reagents" is lower than the detection limit of spectral analysis method, "chromatography reagents" do not have impurity peak with mass lower than 10^{-10} g at the highest sensitivity, "super pure reagents" are used for trace analysis and scientific research. These reagents have special requirements for production, storage and use.

In analytical experiment, the grade of reagents used should accord with the experimental methods, water, vessels used, etc. Generally, solvents used in analytical experiment are analytical reagents (AR) and prepared using distilled or deionized water. In some demanding experiments, if guaranteed reagents (GR) are used, distilled or deionized water cannot be used. In some special circumstances, if the purity of commercial reagents cannot meet the experimental requirements, they should be prepared by the experimenters themselves.

2. Access Principles for Chemical Reagents

The actual purity and grade of chemical reagents used in the analysis work should be comparable to the analytical method. Under the premise of meeting the requirements of the experiment, we must pay attention to the principle of saving.

1) Never Stain Reagents

Chemical reagents cannot be touched by hands. Solids should be taken with a clean spoon. The removed cap should be placed on the experiment table and covered immediately after use to prevent contamination and deterioration. Caps of reagent bottles should not be confused.

2) Save Reagents

In experiments, the amount of chemical reagents should be used according the needs. If it is not specified, it should be used as few as possible. The superfluous reagents cannot be put back to the original reagent bottle to avoid staining.

3) Reagent Access Principle

Chemical analysis experiments usually use analytical pure reagents; instrumental analysis experiments generally use superior pure, analytical pure or proprietary reagents.

3. Storage of Reagents

In the laboratory, the storage of reagents is a very important task. Poor storage may cause deterioration of some reagents, not only leading to the waste of reagents, but also failure of analysis and even accidents. Normally, chemical reagents should be stored in a well-ventilated, clean, dry place to prevent contamination by moisture, dust and other substances. At the same time, different methods of storage are also applied to the reagents with special properties.

(1) Reagents that are easy to corrode glass (fluoride, caustic alkali, etc.) should be stored in plastic bottles or glass bottles coated with paraffin.

(2) Reagents that are easy to be oxidized (stannous chloride, ferrous salt, etc.), weathered or deliquesced ($AlCl_3$, anhydrous Na_2CO_3, NaOH, etc.) should be stored using paraffin to seal the bottle neck.

(3) Reagents that are easy to decompose under light irradiation ($KMnO_4$, $AgNO_3$, etc.) should be stored in brown bottles in the dark.

(4) Reagents that are easy to decompose under heat, reagents with low boiling point and volatile reagents should be stored at cool place.

(5) Reagents with strong water absorbing capacity (anhydrous Na_2CO_3, caustic soda, etc.) should be stored with seal.

(6) Reagents that can react with each other (oxidant and reductant, etc.) should be kept apart.

(7) Reagents with high toxicity (arsenic trioxide, mercury chloride, etc.) must be kept well and used safely.

Chapter 2　Basic Operation for Quantitative Analysis

2.1　Analytical Balance

2.1.1　Weighing Method

The appropriate weighing method must be adopted according to different objects and requirements. There are three weighing methods commonly used.

1. Direct Weighing Method

After setting the zero point of the balance, place the weighing object on the weighing pan of the analytical balance. After the balance reaches equilibrium, the reading is the mass of the weighing object. It is required that the object is stable in the air and not easy to absorb water, such as metal, mineral samples, small beakers, etc.

2. Fixed Mass Weighing Method

This method is used for a fixed mass. The sample is required to be non-hygroscopic in air, such as a powder of metal, alloy or small particles. Firstly, weigh the mass of the sample vessel by direct weighing method, then press TAR key and gradually add the sample to the sample vessel with a small spoon until the balance reaches equilibrium and the displayed data is consistent with the mass of the material to be weighed.

3. Decrement Method

This method requires a certain range of the quality of the weighed sample. It is often used to weigh substances that are hygroscopic, easily oxidized or reacted with CO_2. Firstly, put a clean small paper strip on the weighing bottle [Fig.2.1(a)] and take the weighing bottle containing the sample from the dryer to the weighing pan of the analytical balance (be careful not to let your fingers touch any part of the weighing bottle), then press TAR key. Take out the weighing bottle, place it on top of the sample container, clamp the cap with a small piece of paper, open the cap, tilt the weighing bottle and gently tap the upper part of the weighing bottle with the cap to make the sample slowly falling into the container, as shown in Fig. 2.1 (b). When the poured sample is close to the required mass, slowly raise the bottle, and then tap the upper part of the bottle with the cap to make the sample sticking to the bottle neck back into the bottle, cover the stopper, and then put the weighing bottle back on the weighing pan and weigh it. The absolute value displayed by the balance is the mass of the sample.

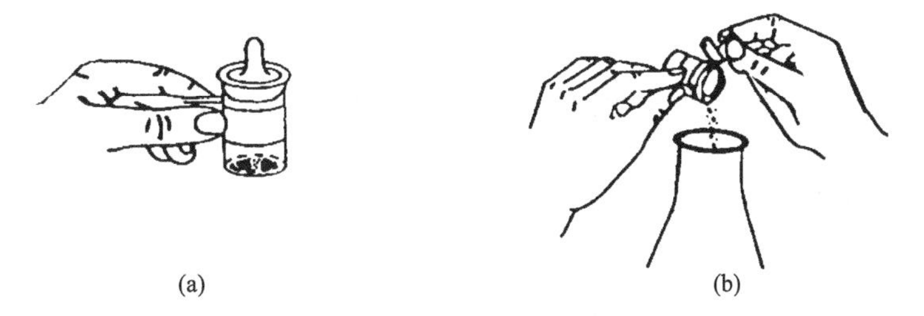

(a) (b)

Fig. 2.1 Decrement method

2.1.2 Operation of Analytical Balance

(1) The cover of analytical balance is folded and placed in a drawer before weighing. The container holding the sample is placed on the left side of the balance.

(2) Check the level of the analytical balance. If the small bubble of the level is offset, adjust the leveling foot until the small bubble is inside the circle of the level.

(3) Turn on the balance, the display is on, and the balance performs a self-test. After about 3 s, the balance model is displayed and then enters the weighing mode, such as 0.0000 g. Newly installed balances or long-term unused balances should be calibrated before use, usually by TAR (or clear) key, CAL key and corresponding standardized weight.

(4) Press the TAR key, display 0.0000 g, and weigh it according to the desired weighing method. The side door should be closed during weighing, and the weighing data should be recorded in the special record book in time, not on a paper or other place. In the same experiment, the same analytical balance should be used to reduce the error.

(5) After weighing, take out the weighing object, close the side door and check the zero point. Close the analytical balance, clean its inside and outside. Then register the usage in the registration book. Cut off the power, and finally cover the balance and put the seat back.

Notes

(1) The load of the analytical balance cannot exceed its maximum load.

(2) The weighing object must be consistent with the temperature inside the balance. Do not put hot or cold objects into the analytical balance for weighing. In order to prevent moisture, the desiccant for moisture absorption, such as a color changing silica gel, should be placed in the balance.

(3) The front door of the analytical balance should not be opened at will. It is mainly used for installation, debugging and maintenance of the analytical balance.

(4) Chemical reagents and samples should not be placed directly on the weighing pan. They should be placed in the clean watch glass, weighing bottles or crucibles. Corrosive gas or hygroscopic substances must be placed in the weighing bottles or other closed containers.

2.2 Pipetting and Constant Volume Operation

2.2.1 Volumetric Flask

Volumetric flask is an input type of the graduated vessel that can accurately measure the volume of the contained solution. It is a thin-necked pear-shaped flat-bottomed glass bottle with a ground glass stopper or plastic stopper. There is a marking line on the neck, which generally refers to the volume of the solution when it is filled until the lower edge of the meniscus is tangent to the marking line at the indicated temperature. Commonly used volumetric flasks are available in 10 mL, 25 mL, 50 mL, 100 mL, 250 mL, 500 mL, and 1000 mL.

The volumetric flask is mainly used to prepare a standard solution of accurate concentration or to dilute the solution quantitatively. Check the volumetric flask before using to ensure no leakage and the position of the marking line is not too close to the bottle mouth. If the water leaks, the solution cannot be accurately prepared; if the marking line is too close to the bottle mouth, it is inconvenient to mix the solution, so it is not suitable for use. For checking leakage, add tap water to the vicinity of the marking line, cover the bottle stopper, pinch the upper part of the marking line on bottleneck by left hand, hold the bottle stopper with the index finger, and hold the bottle edge with the tip of the right finger. Stand upside down for 2 min. If there is no water leakage, set the bottle upright, turn the stopper 180°, and then stand upside down for 2 min. If it does not leak, it can be used. When using the volumetric flask, to avoid contaminated or leakage, a rubber band can be used to tie the bottle stopper to the bottleneck. If the stopper is a flat plastic stopper, the stopper should be inverted on the table. Volumetric flask should be cleaned before use after the leakage detection.

Preparation of solution: accurately weigh the solid substance (reference reagent or test sample) in a small beaker, and dissolve with appropriate amount of water. Then the solution is quantitatively transferred to a volumetric flask. The method of transferring the solution is shown in Fig. 2.2. Hold the glass rod by one hand and stretch into the volumetric bottle mouth. The lower end of the rod rests below the marking line of the inner wall of the bottleneck. Hold the beaker by another hand and let the beaker mouth close to the glass rod. Slowly tilt the beaker, make the solution flow along the glass rod to the volumetric flask. After pouring the solution, raise the beaker slightly along the glass rod, and erect the beaker to return the droplets between the glass rod and the beaker mouth to the beaker. The glass rod is placed back into the beaker, the beaker and the glass rod are then washed 3～4 times with a small amount of pure water by a washing bottle, and the washing liquid was all transferred into the volumetric flask. Dilute with pure water to 2/3 of the volume of the volumetric flask, and shake the volumetric flask in the same direction to mix. Do not reverse the volumetric flask at this time. Continue to add water to about 1 cm below the marking line, wait 1～2 min, let the solution flow down the inner wall of the bottleneck, and finally add pure water with a dropper or washing bottle until the lower edge of the meniscus is tangent to the marking line. Cover the dry stopper, pinch the upper part of the marking line on bottleneck by left hand, hold the stopper with the index finger, hold the bottom of the bottle with the tip of the right finger, turn the volumetric flask upside down and shake it, then turn it up, so that the bubble rises to the top. Repeat the procedure many times to mix the solution thoroughly, as shown in Fig. 2.3.

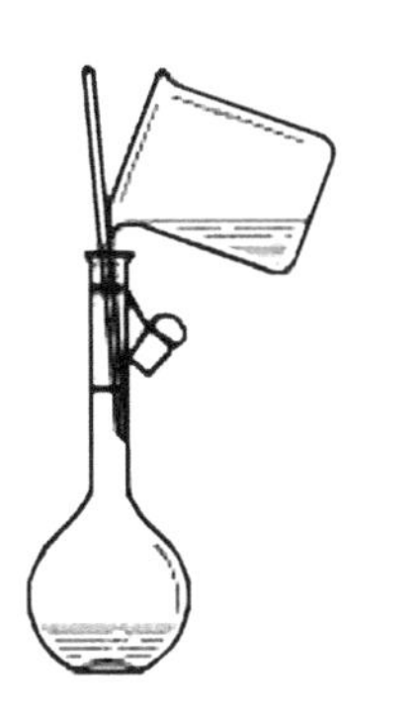

Fig. 2.2 Quantitative transfer of solution

Fig. 2.3 Mix solution

Dilution of solution: a certain volume of solution is transferred into the volumetric flask by the pipette. And then perform the above procedure as mentioned.

Notes

(1) Hot solution should be cooled to room temperature before dilution, otherwise it will cause the volume error.

(2) Solution sensitive to light should be prepared using a brown volumetric flask.

(3) Volumetric flask should not be used to store solution for a long time. The prepared solution should be stored in a reagent bottle, which is rinsed with prepared solution 2～3 times.

(4) Volumetric flask should be rinsed timely after using. Its grinding mouth should be washed, dried and separated by paper if the volumetric flask will not be used for a long time.

2.2.2 Pipette

Pipette is an output type of the graduated vessel that can transfer a certain volume of solution. One type of pipette is a slender, intermediate-expanded glass tube with a circular marking line on the upper part of the neck. The bulk is marked with its volume and the temperature of the standardization. At the indicated temperature, draw the solution until the lower edge of the meniscus is tangent to the marking line and let the solution flow freely in a certain way. The volume of the solution is equal to the volume indicated on the tube. Another type of pipette is a glass tube with the graduation lines, which is typically used to measure a smaller volume of solution. The graduation lines may be engraved to the tip of the pipette or 1～2 cm away from the tip of the pipette. Commonly used pipettes are available in various sizes such as 1 mL, 2 mL, 5 mL, 10 mL, 25 mL and 50 mL (Fig. 2.4).

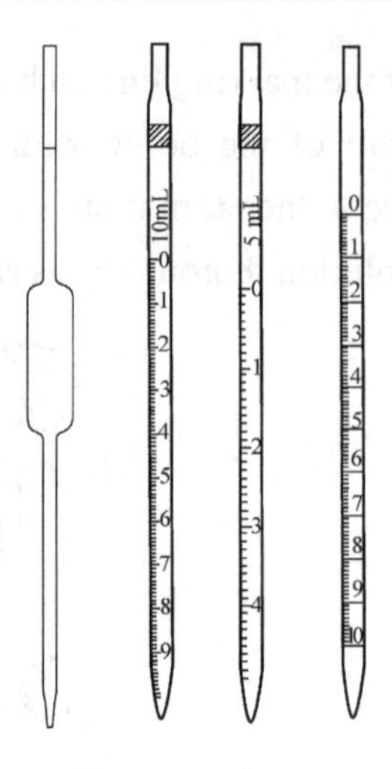

Fig. 2.4　Pipette

1. Rinse of Pipette

Generally, use washing ear ball to absorb chromic acid lotion to wash the pipette. It can also be soaked in a cleaning solution, take out it and drain the cleaning solution, then flush with tap water and distilled water. Water must flow from the tip of pipette.

2. Use of Pipette

Remove the water inside and outside of the pipette tip by filter paper before transferring solution, and then rinse the pipette 3 times with the solution. While transfering the solution with pipette, hold the pipette using thumb and middle finger of right hand, insert the pipette into the solution 1～2 cm lower than the surface. If it is too deep, more solution is adsorbed; if it is too shallow, air is easy to get in. Hold the washing ear ball using left hand and force out the air, put the tip end of the washing ear ball into the mouth of pipette, and loose it to absorb the solution, as shown in Fig. 2.5 (a). The pipette should be descended with the surface of solution. When the liquid level of solution in the pipette is higher than the marker, remove the washing ear ball. Press the mouth of pipette using the index finger of right hand, lift the pipette, and make the pipette two turns against the inner wall of the container to remove the solution adsorbed on the tip of pipette. Tilt the container (30°) and erect the pipette. Close the tip of pipette to the inner wall of the container. Slightly relax the index finger and gently rotate the pipette using thumb and middle finger to let the solution flow out smoothly along the wall. Until the lowest point of the lower edge of the meniscus is tangent to the marking, immediately press the mouth of pipette with index finger to prevent the solution from flowing out. Move the pipette to the receiving container, and close its tip to the inner wall of the container. The pipette should be vertical, and the container should be tilted at about 30°. Loosen the index finger to allow the solution to flow freely along the wall of the container, as shown in Fig. 2.5 (b). After the solution is completely discharged, wait for another 15 s, and then take out the pipette. While using scale pipette, make the liquid level down steadily, after the solution is pouring to the volume required, the index finger is pressed tight, and remove the pipette. If the tube is not marked with "blow", do not blow the solution remaining in the tip of the tube into the receiving container, because the volume of the solution retained at the tip has been taken into account when calibrating the pipette.

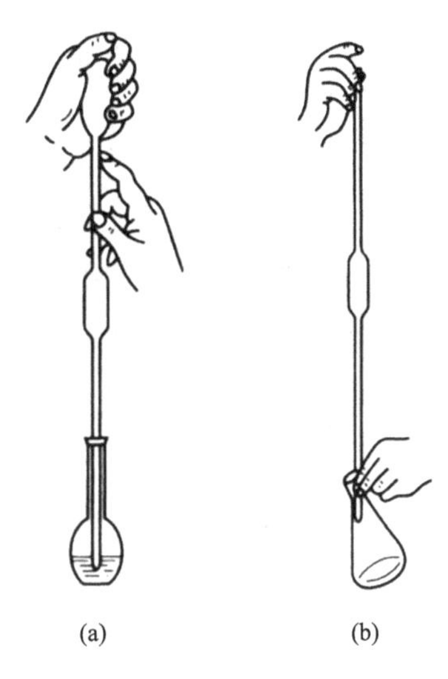

(a) (b)

Fig. 2.5 Use of pipette

2.3 Burette and Basic Operations for Titration Analysis

Burette is a graduated vessel used to accurately measure the volume of the standard solution flowing out during titration. According to the function, it is divided into acid burette and basic burette (Fig. 2.6). There is a glass stopcock at the lower end of the acid burette for acidic, neutral and oxidizing solutions. It is not suitable for alkaline solution (to avoid corrosion of the grinding and stopcock). The lower end of the basic burette is connected with a latex tube with glass bead to control the flow rate of the solution, the lower end of the latex tube is connected with a tip glass tube, which is mainly used for alkaline and non-oxidizing solutions. The solution that can react with latex tubes, such as $KMnO_4$, I_2, silver nitrate, etc., should not be contained in the basic burette. There is also a universal burette with a PTFE stopcock on the market. This burette can overcome the shortcomings of the above-mentioned acid and basic burettes, such as easy clogging of the stopcock, easy aging of the latex tube and only for loading of certain solutions. It is more convenient to use.

Commonly used burettes have nominal capacities of 50 mL and 25 mL, as well as semi-micro or micro-burettes with marked capacities of 10 mL, 5 mL, 2 mL, and 1 mL. The burette with a nominal capacity of 50 mL has a minimum scale of 0.1 mL and a reading accuracy of 0.01 mL.

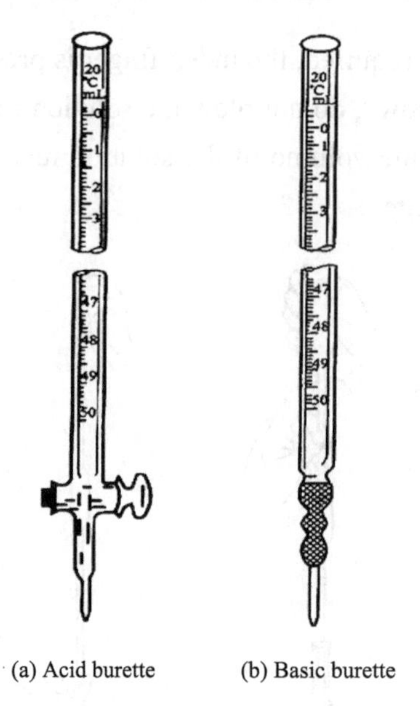

(a) Acid burette (b) Basic burette

Fig. 2.6 Burette

2.3.1 Preparation of the Burette before Use

Before using the acid burette, check if the glass stopcock is flexible and then check the leakage. The method of leakage test is: first turn off the stopcock, fill the burette with water, clamp the burette on the burette shelf for 2 min to check leakage. Rotate the stopcock 180° to ensure no leaking. Otherwise, take out the stopcock and wipe up, and spread on some Vaseline (for gas seal and lubrication) before use.

The method of spreading on Vaseline is shown in Fig. 2.7. Drain water in the burette, place it on the experiment table, take out the stopcock, and wipe up it using absorbent paper. Spread on a thin layer of Vaseline. Be careful not to block the hole of stopcock. Then insert the stopcock and press. Rotate the stopcock in a same direction until the Vaseline on the stopcock is uniform and transparent. After above treatment, the stopcock should be rotate flexibly and the Vaseline layer does not have trace.

Fig.2.7 Method of spreading on Vaseline

Right size of glass bead and latex tube should be chosen for basic burette. If the glass bead is too small, it is easy to slide up and down to leak water. If it is too big, it is difficult to drop solution.

The burette should be cleaned before use. The specific method is shown in 1.2.3.

2.3.2 Loading of Standard Solution

Firstly, rinse the burette 2～3 times with 5～10 mL of standard solution. During operation, the flat-end burette is slowly rotated to allow the standard solution to flow through the entire tube and allow the solution to flow out from the lower end of the burette to remove residual moisture from the tube. The standard solution after mixing should be directly poured into the burette. Do not use any other vessels to avoid dilution or contamination. After the standard solution is loaded, check whether there is any bubble in the tip of the burette, otherwise the bubble will affect the accuracy of the measured volume during the titration. For the acid burette, rotate the stopcock to allow the solution to quickly rush out and carry the bubbles away. For the basic burette, hold the upper end of the burette with right hand and tilt the tube. Squeeze the latex tube around the glass beads with left hand, and tilt the tip to spray the solution from the tip to remove the bubbles (Fig. 2.8). Add standard solution above the scale of "0" and adjust the liquid level to 0.00 mL or a little below, and then record the volume after 0.5～1 min.

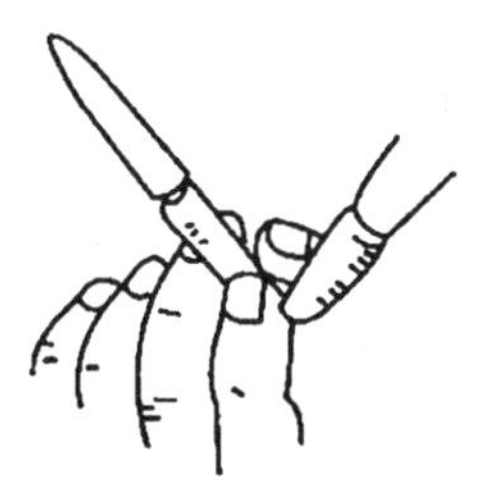

Fig. 2.8 Method of removing air bubbles in basic burette

2.3.3 Burette Reading

Generally, inaccurate reading of the burette is one of the main sources of error in the titration. Therefore, the following rules should be obeyed.

(1) When the solution in the burette is filled or let out, wait 1～2 min to drop the solution adsorbed on the inner wall of burette and then begin to record the volume. If the flowing rate is slow or near the endpoint, wait 0.5～1 min. Before recording the volume every time, check whether there is droplet and air bubble in the tip of burette.

(2) Taken down the burette from the shelf when reading. Hold the burette using thumb and index finger and keep it vertical.

(3) For a colorless or light solution, always keep your eyes at the same level of the lowest point of the lower edge of the meniscus. The initial and final reading should be read by the same method.

(4) If blue ribbon burette is used, the liquid surface represents a triangle intersection point. The reading should be at the cross-point.

(5) The liquid level should be adjusted to 0.00 mL or slightly below before each titration to

reduce error.

(6) The reading should be reached the second number after the decimal point and should be accurate to 0. 01 mL.

(7) In order to make the reading more accurate, a reading card can be used to help the beginners to practice reading. Put the reading card behind the burette, read the scale of the black part tangent to the lower edge of the meniscus of the solution. The reading method should be the same, which means using a reading card all the time or not.

2.3.4 Titration Operation

During the titration, the burette should be directly clamped on the burette shelf, and the titration station should be white. Otherwise, a white porcelain plate should be placed as a background, which is convenient for the observation of the color change of the solution. The titration is preferably carried out in an Erlenmeyer flask and, if necessary, in a beaker. The titration operation is shown in Fig. 2.9.

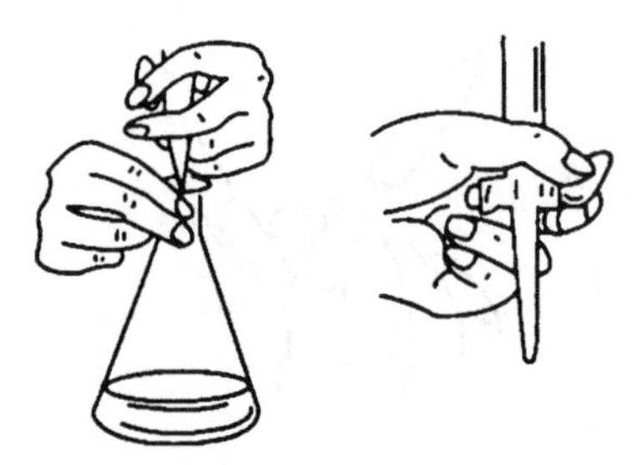

Fig. 2.9　Titration operation of acid burette

When using an acid burette, control the stopcock by left hand. The thumb is at the front, and the index finger and middle finger are behind. The fingers slightly bent to hold inward the stopcock. When rotating the stopcock, take care not to touch the end of stopcock to avoid loosing and leaking of solution. Hold the Erlenmeyer flask with right hand. It is appropriate to stretch a little of burette tip into the flask. In the titration, keep shaking to blend the solution and complete reaction. In the titration, left hand should not leave the stopcock and the change of color should be observed carefully. When the endpoint is near, the titration speed should be slowed down, shake the Erlenmeyer flask after each drop, and then use a washing bottle to blow a small amount of pure water to rinse the inner wall of the Erlenmeyer flask.

When using a basic burette, squeeze the latex tube around the glass bead by the thumb and index finger of left hand. Then the solution drops from the gap between the latex tube and glass bead. The glass bead should not be squeezed forcedly or moved up and down. After titration, loose the thumb and index finger and then the other three fingers.

2.4　Basic Operations of Gravimetric Analysis

The basic operations of gravimetric analysis include sample dissolution, precipitation, filtration, washing, drying (or burning), weighing etc. The correct operation of any process will affect the final analysis results, so each step must be careful and accurate.

2.4.1 Sample Dissolution

There are two methods for sample dissolution, one is to dissolve with distilled water or acid, and the other is to dissolve in solvent after high temperature melting.

2.4.2 Precipitation

The conditions for crystal precipitation should obey the following principles.

Dilute: the precipitated solution should be properly diluted.

Heat: the solution should be heated during precipitation.

Slow: the rate of addition of the precipitant is slow.

Stir: stir constantly with a glass rod during precipitation.

Aged: after the precipitation is complete, it should be aged for a while.

In order to achieve the above requirements, hold the dropper in one hand and slowly add the precipitant during the sedimentation operation; hold the glass rod in another hand to continuously stir the solution. The glass rod should not touch the inner wall and the bottom of the beaker during the stirring process, and the speed should not be fast, so as to avoid the solution spilling. The heating should be carried out on a water bath or a hot plate. The solution should not be boiled, otherwise it will cause water splashing or foaming, resulting in loss of the analyte. After the precipitation is completed, the sediment should be checked for completeness by holding the precipitation solution for a while, allowing the sediment to sink. After the supernatant solution is clarified, a drop of precipitant is added to observe whether the interface is turbid. If it is turbid, the precipitation is not complete and more precipitant needs to be added; otherwise, the precipitation is complete.

After the precipitation is complete, cover the watch glass, and leave it for a period of time or keep it on water bath for 1 h, let the precipitated small crystals form large crystals, and the incomplete crystals turn into complete crystals.

2.4.3 Filtration and Washing

The purpose of filtration and washing is to separate the precipitate from the mother liquid, the excess precipitant and other impurity components, and convert the precipitate to a neat, single component by washing. For the precipitate that needs to be burned, it is often filtered and washed with filter paper in a glass funnel; and the precipitate which can be weighed only by drying is filtered and washed in a crucible. Filtration and washing must be done at one time without interruption. No loss of sediment during operation.

1. Filter Paper

Filter paper is divided into two categories: qualitative and quantitative filter paper.

Quantitative filter paper is used in gravimetric analysis. After the quantitative filter paper is burned, the ash content is less than 0.0001 g, which is called “ashless filter paper”, and its quality is negligible. If the ash mass is greater than 0.0002 g, the mass is deducted from the precipitate. Generally, the ash content of commercially available quantitative filter paper has been marked for reference. Quantitative filter paper is generally round, divided into three types according to the pore size: fast, medium and slow speed. It should be selected according to the nature of the sediment when filtering. For example, for the crystal precipitation, a slow filter paper with a small pore size should be used; and for an amorphous precipitate, a fast filter paper with a large pore size should be used. The size of the filter paper should be determined according to the amount of sediment, and the height of the sediment does not exceed 1/3 of the height of the filter cone.

Fold the filter paper: fold the filter paper in half, and then fold it in half again to form a right angle, and expand it into a cone. Make one layer on one side and three layers on the other side. Place it in a clean funnel (the upper edge of the filter paper should be slightly lower than the upper edge of the funnel). If it is not completely tight between the funnel and the filter paper, adjust the folding angle of the filter paper until it is completely closed. In order to make the filter paper adhere to the inner wall of the funnel without bubbles, a small amount of the three-layer outer corners can be torn off and stored on a clean and dry watch glass for later use.

Place the filter paper: put the folded filter paper into the funnel, the three-layer should be on the short side of the funnel neck, press the thick side with finger, blow a small amount of water to wet the filter paper, then gently press the filter paper to remove the bubbles, thus there is no gap between the upper portion of the tapered filter paper and the funnel. Add water to the edge of the filter paper, and now the funnel is filled with water to form a water column. When the water in the funnel is completely drained, the water column can still remain in the neck without bubbles. If a complete water column cannot be formed, use a finger to block the funnel outlet, slightly pick up the three-layer thick side of the filter paper, and fill the gap between the filter paper and the funnel with a washing bottle until most of the funnel neck and cone are filled with water. Then press the edge of the filter paper to remove air bubbles. Finally, slowly loosen the finger blocking the outlet of the funnel, and the water column can be formed. The filtration speed can be significantly accelerated by the suction of the water column during the process of filtration and washing. Place the prepared funnel on the funnel holder, and place a clean beaker underneath to take the filtrate. The long side of the funnel neck outlet should be close to the cup wall so that the filtrate flows down the wall to avoid splashing. The height of the funnel is preferably such that the outlet of the funnel does not contact the filtrate when filtering.

2. Filtration with Filter Paper

The filtration is carried out in three steps: the first step is to use the pouring method to filter the supernatant as possible; the second step is to transfer the precipitate to the funnel; and the third step is to clean the precipitate on the beaker and the funnel. This three-step operation must be completed in one operation and cannot be interrupted, especially for gelatinous precipitates.

Filtration by the pouring method, that is, after the sediment is settled, the supernatant liquid is poured into the funnel along the glass rod. Allow the sediment to remain in the beaker as possible, then add the washing solution to the beaker, stir the precipitate to wash it thoroughly, let it stand for clarification, and then pour out the supernatant, which will speed up the filtration, prevent the filter paper from being clogged, and make the precipitate fully washing. During the operation, move the beaker with left hand to the top of the funnel, slowly remove the glass rod from the beaker and lean it against the inner wall of the beaker, so that the droplet hanging from the lower end of the glass rod flows into the beaker. Then slowly tilt the beaker to make the supernatant slowly injected into the funnel along the glass rod. When the liquid level of the solution is poured to about 0.5 cm below the edge of the filter paper, the pouring should be suspended to prevent the precipitation from coming out of the edge of the filter paper due to capillary action. When the pouring is stopped, slowly raise the beaker mouth along the glass rod, stand the beaker upright and put the glass rod back into the beaker to prevent a small amount of deposit on the beaker mouth from sticking to the glass rod. The pouring method is shown in Fig. 2.10.

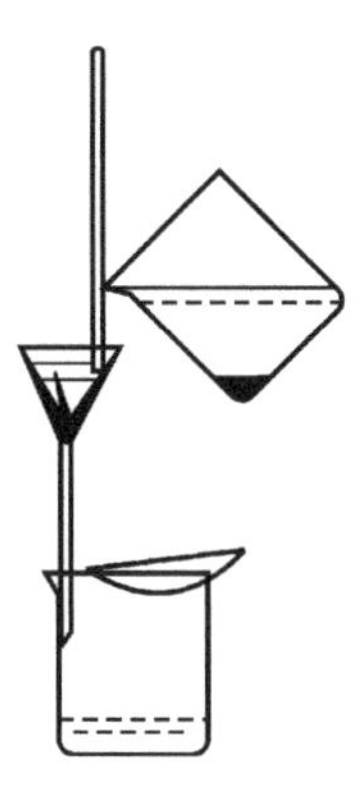

Fig. 2.10 Filtration by the pouring method

After the supernatant is poured, the initial washing can be performed. Spin and wash the inner wall of the beaker and the glass rod from top to bottom with 15～20 mL of washing solution or water from washing bottle or dropper. Stir the precipitate with a glass rod to wash it thoroughly. Then tilt the beaker to sink the sediment on one side to facilitate separation of the sediment and supernatant, facilitating the transfer of the supernatant. After clarification, the mixture was poured, filtrated and washed as the same procedure mentioned above for 3～4 times.

After initial washing, the precipitate can be quantitatively transferred. Add a small amount of washing solution to the beaker containing the precipitate, stir thoroughly with a glass rod, and immediately transfer the suspension to the filter paper. For the last small amount of precipitate remaining in the beaker, repeat the above operation and transfer it completely to the filter paper. After the precipitate is completely transferred to the filter paper, the final washing is performed on the filter paper. A small and slow liquid flow is spirally blown down from the upper portion of the filter paper along the funnel wall to make the precipitate concentrate at the bottom of the filter cone, as shown in Fig. 2.11. After washing several times, about 1 mL of the filtrate was taken with a clean watch glass, and a sensitive and rapid qualitative reaction was selected to check whether the precipitate was well washed.

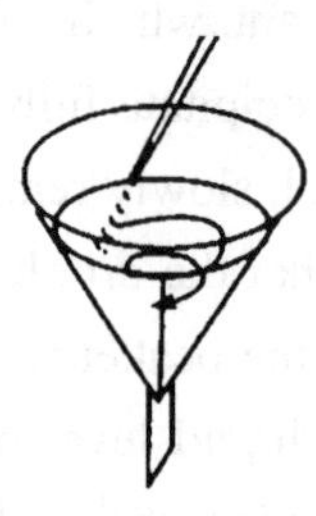

Fig. 2.11　Precipitation washing

3. Filtration with a Micro-porous Glass Funnel or a Micro-porous Glass Crucible

The precipitate which can be weighed after drying or has poor thermal stability should be filtered in a glass filter. Two types of glass filters are shown in Fig. 2.12. The filter plates are made by melting glass powder at high temperature. None of these filters can filter strong alkaline solutions to prevent strong alkali from corroding glass pores. The glass funnel or crucible can be divided into six grades, that is, G1 to G6 (or No. 1 to No. 6) according to the pore size of the micro-pores.

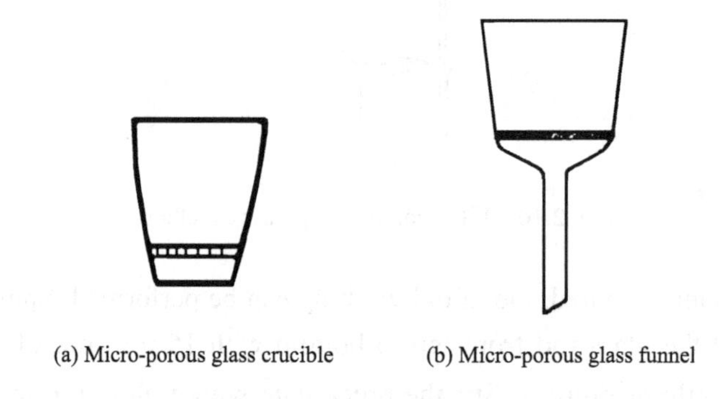

(a) Micro-porous glass crucible　　(b) Micro-porous glass funnel

Fig. 2.12　Micro-porous filter

The micro-porous glass filter must be used by the pouring method under the condition of suction filtration. The operation of filtration, washing and transfer of the sediment are the same as that in the filter paper method.

2.4.4　Drying, Burning and Weighing

The precipitate obtained by filtration is subjected to heat treatment to obtain a precipitate with the constant composition, which is completely identical to its chemical formula.

1. Drying the Precipitate

Drying is generally carried out below 250℃. Any precipitate that has been filtered with a micro-porous glass filter can be treated by a drying method. The method is to place the micro-porous glass filter together with the precipitate on the watch glass and place it in an oven at a suitable temperature. The first drying time can be slightly long (such as 2 h), and then the second drying time can be shortened to 40 min. After the precipitation is dried, it is placed in a dryer and cooled to room temperature and weighed. Repeat the above-mentioned operation several times until the weight is constant. Note that the operation conditions should be consistent.

2. Preparation of Crucible and Use of Dryer

The cobalt or iron salt solution is used to identify the number of the crucible and its lid after washing and drying. Then, the empty crucible is preliminarily burned in a high temperature furnace for 15~30 min to a constant weight under the temperature for burning precipitate. The burnt crucible is naturally cooled and then clamped into a dryer (the use of the dryer is shown in Fig. 2.13). Do not cover the dryer until the hot air escapes. The crucible can be weighed after cooling to room temperature. Then burn for 15~20 min, cool and weigh until the difference between the consecutive two masses is less than 0.2 mg. It can be considered that the crucible has been constant weight.

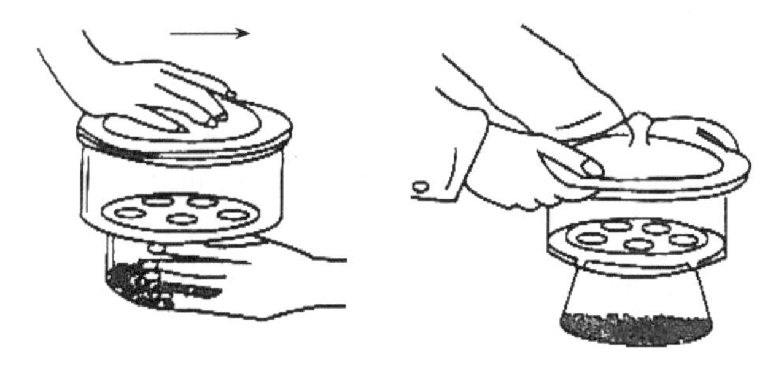

Fig. 2.13 Use of dryer

3. Package of Precipitate

Use a clean spatula or a top round glass rod to pry up two layers of the three-layer of the filter paper, then remove it from the raised filter paper with the clean fingers and open it into a semicircle, fold it to the right from the 1/3 radius of the left end and then to the left from the 1/3 radius of the right end, then fold it from top to bottom and roll it from right to left into a small roll, as shown in Fig. 2.14 (a), and finally put it into the crucible with constant weight. The side with more layers is facing up to facilitate carbonization and ashing. For fluffy sediments such as colloid, use a glass rod to pick up the periphery of the filter paper and fold it inward, then seal the mouth of the cone, as shown in Fig. 2.14 (b). Take the filter paper out and turn it upside down, put it in the crucible with the tip facing up.

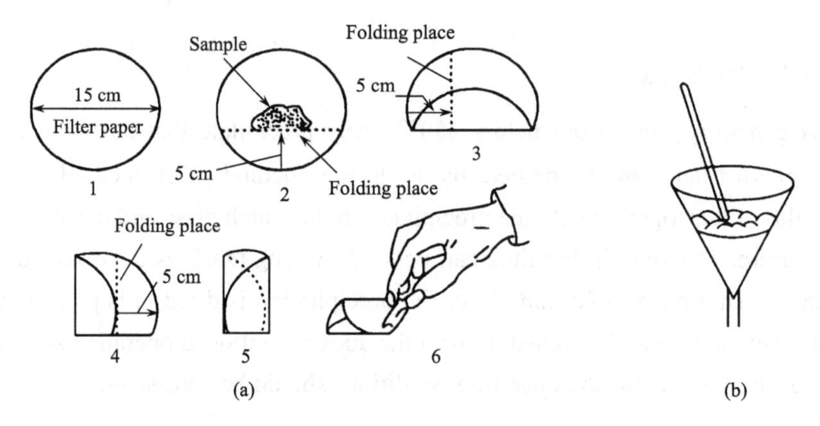

Fig. 2.14 Precipitation package

4. Drying, Burning and Weighing of the Precipitate

Place the crucible with precipitate on the pipeclay triangle, make the multi-layer of filter paper face up, and dry the filter paper and the precipitate until the filter paper is completely carbonized (the filter paper turns black). After carbonization, the temperature can be gradually increased to make the filter paper ashing.

After the precipitate and filter paper are ashed, the crucible is transferred into a high temperature furnace (adjusting the temperature according to the nature of the precipitate), and the lid is covered with a void. Burn it for 40～45 min at the same temperature as burning the empty crucible, take out it and cool to room temperature, and weigh. Then the second and third burnings are carried out until the crucible and the precipitate are constant weight. Usually only 20 min is needed to burn after the second burning. The constant weight means that the weighing difference between consecutive two burnings is less than 0.4 mg. The crucible should be slightly cooled in the air after burning and taking out of the furnace, then transferred to a dryer, cooled to room temperature and weighed. Then burn, cool and weigh it until constant weight is obtained.

Note that the time for burning, weighing, and storage is always consistent every time. After the filter paper is all white, it is moved to a high temperature furnace and burned to a constant weight, and then weighed. The burning conditions for the precipitate in the crucible and the requirement for constant weight should be the same as that for empty crucible.

Chapter 3 Quantitative Analysis Experiments

3.1 Basic Operation Practices

Exp.1 Practices of Analytical Balance

Objectives

(1) To be familiar with the operation of analytical balance.

(2) To learn common weighing methods and be proficient in the decrement method.

Principles

The analytical balance is a necessary precision measuring instrument for quantitative analysis experiments. For details on the use methods and notes of the analytical balance, and the weighing method of a sample, please see Section 2.1. The self-weight of the analytical balance is light, so the use of the analytical balance may cause movement due to collision, which may change the level and affect the accuracy of weighing. Therefore, always be careful, slow and steady during the operation process. Do not use too strong forces to avoid damaging the balance.

Materials

Equipment: analytical balance (0.1 mg), beaker (100 mL), weighing bottle.

Reagents: anhydrous Na_2SO_4 (s).

Procedures

(1) Fixed mass weighing method (weigh 3 parts of 0.5000 g sample).

Turn on the analytical balance and place the clean, dry beaker in the center of the weighing pan after adjusting the zero (use the paper strip when handling the beaker, and operate according to the regulations), close the side door, and then press the clear (or TAR) key. After the balance shows 0.0000 g, open the side door of the balance and slowly add the sample to the center of the beaker with a small spoon until the balance shows 0.5000 g. If the amount weighed is less than this value, the sample may continue to be added; if the amount displayed exceeds this value, it should be re-weighed. Record the data in time after each weighing in Table 3.1.

(2) Decrement method (weigh 3 parts of 0.5000 g sample).

(i) Place the weighing bottle containing the sample in the center of the weighing plate. Press the clear (or TAR) key. The displayed value is 0.0000 g.

(ii) Take out the weighing bottle according to the regular operation, and slowly knock out the sample of the required quality above the small beaker. After the sampling is completed, close the cap, and place the weighing bottle on the weighing plate. The displayed value at this time is the quality of the taken sample (regardless of the negative sign). If the mass is less than the weighing range, continue to knock out part of the sample until its mass is within the required range; if the mass exceeds the upper limit of the weighing range, the reading is not recorded. Press the clear (or TAR) key, after the displayed value is 0.0000 g, weigh it again. Repeat the above operation, after each weighing value is obtained, press clear (or TAR) key, and then carry out the next weighing, so that a series of weighing values can be obtained continuously and easily. Record the data in time after each weighing in Table 3.1.

Data Recording

Table 3.1 Practices of weighing

	1	2	3
Fixed mass weighing method, m/g			
Decrement method, m/g			

Notes

Make sure to use the same analytical balance during the experimental process to eliminate errors.

Questions

(1) Which decimal place should be reached when recording the weighing data? Why?

(2) Why should the weighing object be placed in the center of the weighing plate?

(3) What should be done to prevent the sample from loss when using a weighing bottle?

Exp.2 Practices of Basic Operation for Titration Analysis

Objectives

(1) To grasp the operation of the acid burette and basic burette.

(2) To practice the basic operations for titration and the judgment of the endpoints of the commonly used indicators.

(3) To learn how to accurately read and record the readings of burettes.

Principles

While using the HCl and NaOH solution to titrate each other with a certain indicator, the consumption of the titrant will always change with the different volume of the solution being titrated. But the ratio of the consumed volumes (V_{HCl}/V_{NaOH}) should be essentially unchanged. Thus, when the accurate concentrations of the HCl and NaOH solutions are unknown, calculate the precision of V_{HCl}/V_{NaOH} to test your titration technique and ability to judge the endpoint.

Materials

Equipment: acid burette (50 mL), basic burette (50 mL), reagent bottles (500 mL, one with a rubber or plastic stopper), Erlenmeyer flask (250 mL), beaker (100 mL), graduated cylinder (10 mL, 100 mL), analytical balance.

Reagents: NaOH (AR), concentrated hydrochloric acid (AR, 1.19 g/mL), methyl orange aqueous solution (0.1%), phenolphthalein solution (0.2%, in ethanol).

In quantitative analysis, all the reagents except indicators are analytical pure, and the used water is distilled water or deionized water.

Procedures

(1) Preparation of NaOH solution (0.1 mol/L).

Quickly weigh 2.0 g of sodium hydroxide with analytical balance and put it into a 100 mL breaker, then dissolve with about 50 mL of distilled water with stirring. Transfer the solution into a 500 mL reagent bottle, add about 450 mL of distilled water, insert the rubber stopper and mix well and label.

(2) Preparation of HCl solution (0.1 mol/L).

Obtain 4.2～4.5 mL of the concentrated HCl with 10 mL graduated cylinder and transfer it into a 500 mL reagent bottle containing a certain amount of water. Wash out the graduated cylinder 2～3 times, and then add the washing to the reagent bottle. Dilute to 500 mL with distilled water, insert the stopper, mix well and label. All the operations above should be carried out in the fume hood because the concentrated hydrochloric acid is easy to volatilize.

(3) Operation of the basic burette and judgment of the endpoint.

Check the leakage of the burette, and wash them (including with the distilled water). Then rinse the burettes 3 times with 8～10 mL of the corresponding operation solutions to remove traces of water in their inner walls and tips. Fill the burettes with the titrants. Free air bubbles from their tips, adjust the liquid levels to or a little lower than zero mark, record the initial volume readings estimated to 0.01 mL.

Add a few milliliters of the HCl solution into a 250 mL Erlenmeyer flask from the acid burette, dilute it with about 20 mL of the distilled water, add 2 drops of the phenolphthalein solution as indicator, and shake well. Add NaOH solution into the Erlenmeyer flask drop by drop through the basic burette (especially the operations of adding one or half a drop of the solution with basic burette), observe the color change of the phenolphthalein indicator near the endpoint. The titration should be continued until the color of the solution becomes reddish and keep for 30 s, which indicating the endpoint. Then add a little amount of HCl solution into the same Erlenmeyer flask to make the color of the solution fade out, then continue to titrate with NaOH solution to the endpoint. Repeat the above operations several times until you can easily control the speed of the titration and accurately judge the endpoint. Practice to read the burettes estimated to 0.01 mL. Note that there are no air bubbles in the rubber connector or the tip of the burettes, and read the data after a certain time.

Based on the previous practices, determine the ratio of the consumed volumes of HCl and NaOH solutions (V_{HCl}/V_{NaOH}) by measuring their accumulative volumes. The procedures are as follows: add about 18 mL of HCl solution in an Erlenmeyer flask from the acid burette (after every release of the solution or titration, accurately record the reading), add two drops of phenolphthalein solution, titrate with NaOH solution to the endpoint. On the basis of the previous readings, 2 mL of HCl solution was added in the same Erlenmeyer flask from the acid burette, and make the color of the solution fade out, then titrate with NaOH solution to the endpoint. During the whole process, do not refill these two burettes, and record the accumulated volumes of the solution consumed. Repeat above steps, and complete the third titration. Calculate the ratios of the consumed volumes (V_{HCl}/V_{NaOH}).

(4) Operation of the acid burette and judgment of the endpoint.

Add a few milliliters of NaOH solution into a Erlenmeyer flask from the basic burette, dilute it with about 20 mL of the distilled water, add 1～2 drops of the methyl orange solution (supply more indicator during the titration process because the volume of the solution in Erlenmeyer flask will increase), and then shake well. The color of the solution is yellow. Titrate with HCl solution (especially note the operations of adding one or half drop of the solution with acid burette), observe the color change of methyl orange indicator near the endpoint. The titration should be continued until the color of the solution turns from yellow to orange, indicating the endpoint. Then add 1～2 mL of NaOH solution into the same Erlenmeyer flask, make the color of the solution return to yellow, continue to titrate with HCl solution. Practice the operation of acid burette, learn how to determine the endpoint, and practice reading the burette.

Based on the previous practices, determine the ratio of the consumed volumes of HCl and NaOH solutions (V_{HCl}/V_{NaOH}) by measuring their accumulative volumes. The procedures are as follows: add about 18 mL of the NaOH solution in an Erlenmeyer flask from the basic burette, add 2 drops of the methyl orange solution, and titrate with HCl solution to the endpoint. On the basis of the previous readings, add 2 mL of NaOH solution in the same Erlenmeyer flask from the basic burette, make the color of the solution changes to yellow, then titrate with HCl solution to the endpoint. Do not refill these two burettes during the whole process, and record the accumulated volumes of the solution consumed in Table 3.2. Repeat above steps, and complete the third titration.

Data Recording and Processing

Calculate the ratios of the consumed volumes (V_{HCl}/V_{NaOH}). The relative average deviation $\bar{d}_r \leqslant 0.3\%$, otherwise you should re-titrate.

Table 3.2 Practices of titration analysis

	1	2	3
V_{HCl} (final)/mL			
V_{HCl} (initial)/mL			
V_{HCl} /mL			
V_{NaOH} (final)/mL			
V_{NaOH} (initial)/mL			
V_{NaOH}/mL			
V_{HCl}/V_{NaOH}			
Average of V_{HCl}/V_{NaOH}			
$\lvert d_i \rvert$			
$\bar{d}_r$ /%			

Questions

(1) Why it is important to rinse the inner wall of the burette with related solution before filling? Is it necessary to dry the Erlenmeyer flask before titration? Why?

(2) While using NaOH solution to titrate the acidic solution with phenolphthalein as indicator, it is emphasized that the titration should be continued until the color of the solution becomes reddish and keeps for 30 s. Why? What are the reasons that the pink color of solution will fade out?

3.2 Acid-base Titration Experiments

Exp.3 Preparation and Standardization of NaOH Standard Solution

Objectives

(1) To grasp the preparation and standardization of NaOH standard solution.

(2) To master the operation of basic burette and identification of the endpoint with phenolphthalein as indicator.

Principles

NaOH is easy to absorb moisture and react with CO_2 in air to generate Na_2CO_3, so NaOH solution is prepared at an approximate concentration and then standardized with a primary standard to obtain its exact concentration.

To prepare the NaOH standard solution without Na_2CO_3, NaOH saturated aqueous solution is usually prepared in advance. Because Na_2CO_3 is not dissolved in NaOH saturated solution, NaOH standard solution can be obtained by diluting the supernatant after the precipitation of Na_2CO_3. In addition, the distilled water for preparing NaOH solution must be heated to boiling to remove CO_2.

There are many primary standards for standardization of alkali solution, such as oxalic acid, benzoic acid and potassium hydrogen phthalate ($KHC_8H_4O_4$, KHP), etc. The KHP standard has the advantages of being easy to produce pure product, not easy to absorb water, stable, and having a high molar mass. In this experiment, KHP is employed to standardize the NaOH solution. The following equation can be used to denote the reaction of titration:

$$C_6H_4(COOH)(COOK) + NaOH = C_6H_4(COONa)(COOK) + H_2O$$

The solution is weakly alkaline at stoichiometric point (pH ≈ 9) because the reaction product is a binary weak base. Therefore, phenolphthalein can be used as indicator.

The concentration of the NaOH standard solution can be calculated by the following equation:

$$c_{NaOH} = \frac{m_{KHP} \times 1000}{M_{KHP} \times V_{NaOH}} \qquad (M_{KHP} = 204.22\ g/mol)$$

Materials

Equipment: basic burette (50 mL), Erlenmeyer flask (250 mL), beaker (250 mL), pipette (10 mL), volumetric flask (1000 mL), graduated cylinder (50 mL), analytical balance, etc.

Reagents: potassium hydrogen phthalate (primary standard), sodium hydroxide (AR), phenolphthalein solution (0.2%, in ethanol).

Procedures

(1) Preparation of NaOH solution (0.1 mol/L).

See details in Exp. 2.

(2) Standardization of NaOH solution (0.1 mol/L).

Accurately weigh 0.4～0.5 g of KHP into a 250 mL Erlenmeyer flask，add 20～30 mL of distilled water. Dissolve it by warming, and add 2～3 drops of phenolphthalein solution after cooling. Titrate with NaOH standard solution until the solution is reddish, and it will not fade for 30 s, which is the endpoint. Record the consumed volume of NaOH standard solution in Table 3.3. Repeat the titration 3 times.

Data Recording and Processing

Calculate the concentration of the NaOH solution and the relative average deviation ($\bar{d}_r \leqslant$ 0.3%).

Table 3.3 Standardization of NaOH solution

	1	2	3
m_{KHP}/g			
V_{NaOH}(final)/mL			
V_{NaOH}(initial)/mL			
V_{NaOH}/mL			
c_{NaOH}/(mol/L)			
$\bar{c}_{NaOH}$/(mol/L)			
$\lvert d_i \rvert$			
$\bar{d}_r$ /%			

Questions

(1) How to calculate the mass range of the primary standard (KHP)? What is the effect if the mass of KHP is too much or little?

(2) What kind of balance is used to weigh the solid NaOH and KHP, respectively? Why?

(3) Which volumetric glass should be used to add 20～30 mL of water to dissolve the KHP, graduated cylinder or transfer pipette?

Exp.4 Preparation and Standardization of HCl Solution

Objectives

(1) To grasp the preparation and standardization of HCl solution.

(2) To master the operation of acid burette and identification of the endpoint with methyl orange as indicator.

Principles

Commercially available concentrated hydrochloric acid is colorless aqueous solution of HCl, with a concentration of HCl around 36%～38%. The density of this solution approximates to 1.18 g/L. It does not meet the requirement of direct preparation of standard solution due to its volatility, thus indirect preparation should be used for HCl standard solution.

The commonly used primary standards for HCl solution are sodium carbonate (Na_2CO_3) and borax ($Na_2B_4O_7 \cdot 10\ H_2O$). In this experiment, Na_2CO_3 is employed as the primary standard.

Na_2CO_3 is easy to adsorb moisture in air, so it should be dried at 270～300℃ for 1 h, and then kept in the dryer before use. The reaction of standardization by Na_2CO_3 is as follows.

$$Na_2CO_3 + 2HCl \xlongequal{} 2NaCl + H_2O + CO_2\uparrow$$

The pH of H_2CO_3 saturated solution is 3.9 at the stoichiometric point. Therefore, methyl orange can be used as indicator.

The concentration of the HCl standard solution can be calculated by the following equation:

$$c_{HCl} = \frac{2m_{Na_2CO_3} \times 1000}{M_{Na_2CO_3} \times V_{HCl}} \qquad (M_{Na_2CO_3} = 105.99 \text{ g/mol})$$

Materials

Equipment: acid burette (50 mL), Erlenmeyer flask (250 mL), pipette (25 mL), reagent bottle, volumetric flask (250 mL), beaker, graduated cylinder (50 mL), analytical balance, etc.

Reagents: concentrated hydrochloric acid (36%～38%, 1.18 g/L), anhydrous sodium carbonate (GR), NaOH solution (0.1 mol/L), methyl orange aqueous solution (0.1%).

Procedures

(1) Preparation of HCl solution (0.1 mol/L).

Transfer 9.0 mL of the concentrated hydrochloric acid to a reagent bottle containing a certain amount of water, dilute to 1000 mL with distilled water and mix well.

(2) Standardization of HCl solution (0.1 mol/L).

Accurately weigh 0.10～0.12 g of anhydrous sodium carbonate into a 250 mL Erlenmeyer flask, dissolve with 20～30 mL of distilled water (or accurately weigh 1.0～1.2 g of anhydrous sodium carbonate to a beaker, dissolve with appropriate amount of water. Quantitatively transfer the solution to a 250 mL volumetric flask, dilute to the mark with distilled water and mix well. Transfer 25.00 mL of the diluted solution to an Erlenmeyer flask). Add 1～2 drops of methyl orange solution as indicator, titrate with HCl solution until the color of the solution changes from yellow to orange, indicating the endpoint. Record the consumed volume of the HCl solution in Table 3.4. Then repeat the titration 3 times.

Data Recording and Processing

Calculate the concentration of HCl standard solution, and the relative average deviation ($\bar{d}_r \leqslant$ 0.3%).

Table 3.4 Standardization of HCl solution

	1	2	3
$m_{Na_2CO_3}$/g			
V_{HCl}(final)/mL			
V_{HCl}(initial)/mL			
V_{HCl}/mL			
c_{HCl}/(mol/L)			
$\bar{c}_{HCl}$/(mol/L)			
$\lvert d_i \rvert$			
$\bar{d}_r$/%			

Notes

(1) The weighing speed should be fast because anhydrous sodium carbonate is very easy to adsorb moisture.

(2) Excessive CO_2 in the solution will increase the solution acidity and advance the endpoint. Therefore, the solution should be shaken vigorously to accelerate the decomposition of H_2CO_3, and heated to remove the excessive CO_2 when approaching the endpoint. Then go on the titration after cooling.

Questions

(1) Why cannot prepare the HCl standard solution directly?

(2) What is the effect on the standardization results if the burette is not rinsed with the HCl standard solution?

(3) Is it possible to use phenolphthalein as the indicator for the standardization of HCl by Na_2CO_3?

Exp.5 Determination of Boric Acid in Food Additives

Objectives

(1) To grasp the principle and method for determining the boric acid by acid-base titration.

(2) To further grasp the operation of basic burette and identification of the endpoint with phenolphthalein as indicator.

Principles

It is difficult to directly titrate boric acid (H_3BO_3) using NaOH standard solution because the dissociation constant (K_a) of H_3BO_3 is too small (7.3×10^{-10}). However, the dissociation constant of boric glycerin (K_a=3×10^{-7}), which can be obtained by adding the glycerin into H_3BO_3 solution, is large enough to be titrated with NaOH standard solution. The reactions are as follows:

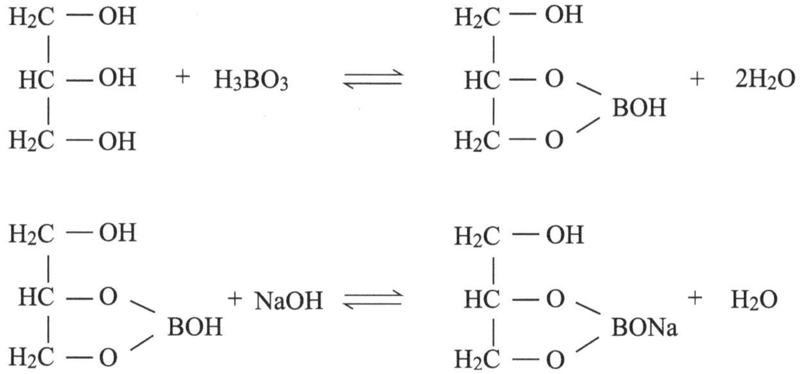

$$\begin{array}{l} H_2C-OH \\ \;| \\ HC-OH \\ \;| \\ H_2C-OH \end{array} + H_3BO_3 \rightleftharpoons \begin{array}{l} H_2C-OH \\ \;| \\ HC-O \diagdown \\ \;| \qquad\quad BOH \\ H_2C-O \diagup \end{array} + 2H_2O$$

$$\begin{array}{l} H_2C-OH \\ \;| \\ HC-O \diagdown \\ \;| \qquad\quad BOH \\ H_2C-O \diagup \end{array} + NaOH \rightleftharpoons \begin{array}{l} H_2C-OH \\ \;| \\ HC-O \diagdown \\ \;| \qquad\quad BONa \\ H_2C-O \diagup \end{array} + H_2O$$

The solution is alkalescent at the stoichiometric point, so phenolphthalein can be used as indicator.

Materials

Equipment: basic burette (50 mL), Erlenmeyer flask (250 mL), beakers, volumetric flask

(250 mL), graduated cylinder (25 mL), analytical balance, etc.

Reagents: potassium hydrogen phthalate (primary standard), sodium hydroxide (AR), neutral glycerin (1：2), phenolphthalein solution (0.2%, in ethanol), boric acid sample.

Procedures

(1) Preparation and standardization of NaOH solution (0.1 mol/L).

See details in Exp. 3.

(2) Determination of boric acid.

Accurately weigh 0.29～0.32 g of boric acid sample in a 250 mL Erlenmeyer flask, add 25 mL of neutral glycerin, dissolve it by gently heating the Erlenmeyer flask, and cool rapidly. Then add 2～3 drops of phenolphthalein solution after the solution is cool. Titrate with NaOH standard solution (0.1 mol/L) until the solution is reddish and the color stands for 30 s, which is the endpoint. Record the consumed volume of NaOH standard solution in Table 3.5. Repeat the titration 3 times.

Data Recording and Processing

According to the recorded experimental data, calculate the concentration of NaOH standard solution and the content of boric acid in sample, as well as their averages and relative average deviations ($\bar{d}_r \leqslant 0.3\%$).

Table 3.5　Determination of boric acid

	1	2	3
$m_{H_3BO_3}$/g			
V_{NaOH} (final)/mL			
V_{NaOH} (initial)/mL			
V_{NaOH}/mL			
$w_{H_3BO_3}$/%			
$\bar{w}_{H_3BO_3}$/%			
$\lvert d_i \rvert$			
$\bar{d}_r$/%			

Notes

(1) Don't shake the Erlenmeyer flask after adding the glycerin, shake it after heating.

(2) At the endpoint, 1～2 drops of phenolphthalein indicator can be supplemented to increase the identification of the color.

Questions

(1) Why cannot determine the boric acid using direct titration method?

(2) What is the conjugate base of the boric acid? Is it possible to directly determine its conjugate base by acid-base titration?

(3) What volumetric glass should be used to measure the glycerin? Is it necessary to rinse this volumetric glass with glycerin? Why?

Exp.6 Analysis of Mixed Alkali

Objectives

(1) To master the principle and method for determining the composition of the mixed alkali by double-indicator technique.

(2) To further acquaint with the operations of acid burette and pipette.

Principles

There are two kinds of the industrial mixed bases: the mixture of NaOH and Na_2CO_3, or the mixture of Na_2CO_3 and $NaHCO_3$. By making use of the HCl standard solution as titrant, meanwhile the phenolphthalein and methyl orange as the indicators, the sample solution can be continuously titrated. According to the volume of the titrant consumed, the composition and content of the mixed alkali can be determined. This method is named as "double-indicator technique".

Add phenolphthalein indicator in the solution, titrate with HCl standard solution until the solution color changes from purple red to reddish, indicating the first endpoint. At this moment, NaOH in the mixed base is neutralized completely, and Na_2CO_3 is titrated to form $NaHCO_3$. The pH of the solution is about 8.32 at the stoichiometric point. Record the consumed volume of the HCl solution as V_1. The titration reactions are as follows:

$$NaOH + HCl = NaCl + H_2O$$

$$Na_2CO_3 + HCl = NaCl + NaHCO_3$$

Add the second indicator (methyl orange) in the same solution, continue to titrate with HCl standard solution until the color of the solution turns from yellow to orange, indicating the second endpoint. At this moment, $NaHCO_3$ react with HCl completely to form NaCl. The pH of the solution is about 3.89 at the stoichiometric point. Record the consumed volume of the HCl solution as V_2. The titration reaction is as follows:

$$NaHCO_3 + HCl = NaCl + H_2O + CO_2$$

According to the values of V_1 and V_2, the composition and content of the mixed alkali can be determined.

If $V_1 > V_2$, the compositions of the sample are NaOH and Na_2CO_3:

$$\rho_{NaOH} = \frac{c_{HCl} \times (V_1 - V_2) \times 10^{-3} \times M_{NaOH}}{25.00 \times 10^{-3}}$$

$$\rho_{Na_2CO_3} = \frac{c_{HCl} \times V_2 \times 10^{-3} \times M_{Na_2CO_3}}{25.00 \times 10^{-3}}$$

If $V_1 < V_2$, the compositions of the sample are Na_2CO_3 and $NaHCO_3$:

$$\rho_{Na_2CO_3} = \frac{c_{HCl} \times V_1 \times 10^{-3} \times M_{Na_2CO_3}}{25.00 \times 10^{-3}}$$

$$\rho_{NaHCO_3} = \frac{c_{HCl} \times (V_2 - V_1) \times 10^{-3} \times M_{NaHCO_3}}{25.00 \times 10^{-3}}$$

If $V_1 = V_2$, the sample only contains Na_2CO_3.

While determining the total alkali content of the industrial mixed base, only add the methyl orange indicator, and titrate with HCl standard solution to the endpoint. The total volume of the consumed HCl solution is V_1 plus V_2. The total alkali content should be calculated by converting the mixed alkali to Na_2O.

Materials

Equipment: acid burette (50 mL), Erlenmeyer flask (250 mL), pipette (25 mL), reagent bottle, volumetric flask (250 mL), beaker, graduated cylinder, analytical balance, etc.

Reagents: HCl standard solution (0.1 mol/L), anhydrous sodium carbonate (primary standard, dried at 270～300℃ for 1 h, and kept in the dryer), phenolphthalein solution (0.2%, in ethanol), methyl orange aqueous solution (0.1%).

Procedures

(1) Preparation and standardization of HCl solution(0.1 mol/L) .

See details in Exp. 4.

(2) Determination of the mixed alkali (double-indicator method).

Transfer 25.00 mL of the alkaline solution to a 250 mL Erlenmeyer flask. Add 2～3 drops of the phenolphthalein solution, titrate with HCl standard solution until the color of the solution turns from purple red to reddish, indicating the first endpoint. Record the consumed volume of HCl solution as V_1 in Table 3.6.

Add methyl orange in the same solution (the color of the solution is slightly orange), continue to titrate with HCl standard solution until the color of the solution turns from yellow to orange, indicating the second endpoint. Record the consumed volume of HCl solution as V_2 in Table 3.6.

Repeat the above titration 3 times.

Data Recording and Processing

(1) Determine the composition of the mixed alkali according to the values of V_1 and V_2.

(2) Determine the content of each component based on the volume of the consumed HCl standard solution. Calculate the mass concentration ρ (g/L) of each component, as well as their averages and relative average deviations ($\bar{d}_r \leqslant 0.3\%$).

(3) According to the values of ($V_1 + V_2$), calculate the mass concentration of Na_2O (g/L) as the total alkali content, as well as their average and relative average deviation ($\bar{d}_r \leqslant 0.3\%$).

Table 3.6 Determination of mixed alkali components

	1	2	3
V_1(final)/mL			
V_1(initial)/mL			
V_1/mL			
V_2(final)/mL			
V_2(initial)/mL			
V_2/mL			
ρ_1/(g/L)			
$\bar{\rho}_1$ /(g/L)			
$\bar{d}_r$ /%			
ρ_2/(g/L)			
$\bar{\rho}_2$ /(g/L)			
$\bar{d}_r$ /%			
(V_1+V_2)/mL			
ρ_{tot}/(g/L)			
$\bar{\rho}_{tot}$ /(g/L)			
$\bar{d}_r$ /%			

Notes

(1) If the titration speed is too fast and the solution is not shaken well near the first endpoint, concentration of HCl in local solution might be too high, and a small amount of Na_2CO_3 would directly react with HCl and release CO_2, resulting in the measurement errors.

(2) Near the second endpoint, be sure to shake the test solution sufficiently to avoid the formation of the supersaturated H_2CO_3 solution, which would result in the increased acidity and the premature endpoint.

Questions

(1) What is the principle and method for the determination of mixed alkali by double-indicator technique?

(2) To determine the composition of the mixed alkali in the following cases:

① $V_1 = 0$, $V_2 > 0$; ② $V_1 > 0$, $V_2 = 0$; ③ $V_1 > V_2$; ④ $V_1 < V_2$; ⑤ $V_1 = V_2$.

Exp.7 Determination of Nitrogen in Ammonium Salts

Objectives

(1) To master the principle of the determination of nitrogen content in ammonium nitrogen fertilizer by using formaldehyde.

(2) To further grasp the operation of basic burette.

Principles

NH_4Cl and $(NH_4)_2SO_4$ are commonly used nitrogen fertilizers, which are strong acid and weak alkali salt. The NaOH standard solution can not be used directly and accurately to titrate NH_4^+ because the acidity of NH_4^+ in the ammonium salt is too weak (K_a=5.6×10^{-10}). The formaldehyde method can strengthen the weak acid, and the reaction is as follows:

$$4NH_4^+ + 6HCHO = (CH_2)_6N_4H^+ + 3H^+ + 6H_2O$$

The resulting mixed acid (K_a of hexamethylene tetramine is 7.1×10^{-6}) can be titrated with NaOH standard solution. Phenolphthalein is used as indicator because the product $(CH_2)_6N_4$ is a weak base. The method of formaldehyde is widely used for the determination of nitrogen content in ammonium salt because it is simple and fast, but its accuracy is less than that of distillation.

Materials

Equipment: basic burette (50 mL), pipette (25 mL), Erlenmeyer flask (250 mL), beaker, analytical balance, etc.

Reagents: potassium hydrogen phthalate (KHP, GR, dried at 105～110℃ to constant weight, and kept in the dryer), NaOH (AR), formaldehyde (40%, AR), phenolphthalein solution (0.2%, in ethanol), $(NH_4)_2SO_4$ sample.

Procedures

(1) Preparation and standardization of NaOH solution (0.1 mol/L).

See details in Exp. 3.

(2) Neutralization of formaldehyde.

Formaldehyde often contains a small amount of formic acid because it is easily to be oxidized. The formic acid should be removed before titration. Take the original supernatant of formaldehyde (40%) in a beaker, add 2 drops of phenolphthalein as indicator, then neutralize with NaOH solution until the formaldehyde solution is reddish.

(3) Sample analysis.

Accurately weigh 0.13～0.16 g (can be estimated based on the exact concentration of the NaOH solution) of ammonium sulfate sample in a 250 mL Erlenmeyer flask, dissolve with 20～30 mL of distilled water, then add 5 mL of neutralized formaldehyde solution and 2 drops of phenolphthalein indicator, shake well. Let it stand for 1 min to complete the reaction, then titrate with NaOH standard solution to slightly reddish and the color does not fade for 30 s. Record the consumed volume of NaOH solution in Table 3.7. Repeat the titration 3 times.

In order to reduce the weighing error, accurately weigh 1.6～1.8 g of ammonium sulfate sample in a beaker and dissolve with about 40 mL distilled water, then transfer to a 250 mL volumetric flask, dilute to the mark with distilled water and shake well. Transfer 25.00 mL of the test solution to a 250 mL Erlenmeyer flask. The following procedures are the same as mentioned above.

Data Recording and Processing

According to the recorded values, calculate the concentration of NaOH solution and the mass content of nitrogen in $(NH_4)_2SO_4$ sample, as well as their averages and relative average deviation ($\bar{d}_r \leqslant 0.3\%$).

Table 3.7 Determination of nitrogen in ammonium sulfate

	1	2	3
m_s/g			
V_{NaOH}(final)/mL			
V_{NaOH}(initial)/mL			
V_{NaOH}/mL			
ρ_N/(g/L)			
$\bar{\rho}_N$/(g/L)			
$\lvert d_i \rvert$			
$\bar{d}_r$/%			

Questions

(1) Can the content of nitrogen in other ammonium salts such as NH_4Cl, NH_4NO_3 and NH_4HCO_3 be determined by formaldehyde method? Why?

(2) What's the weighing error, while weighing 0.13～0.15 g of ammonium sulfate sample in this experiment? What should we do to make the relative error of weighing less than 0.1%?

Exp.8 Determination of Total Phosphorus in Phosphate Fertilizer

Objectives

To master the principle and method for determining total phosphorus in phosphate fertilizer with phosphomolybdic acid quinoline volumetric method.

Principles

The phosphorus in the phosphate fertilizer is converted to phosphoric acid under acidic conditions. The phosphoric acid can react with excess sodium molybdate and quinoline to generate a yellow precipitate of phosphomolybdic acid quinolone under boiling in the nitric acid solution. The precipitate is recovered by filtration and purified after washing. After that, the precipitate was dissolved in a known and excess amount of the standard alkaline solution, and then re-titrate the excess of alkali with acid standard solution. The percentage of phosphorus pentoxide can be calculated based on the consumption of alkali and acid in titration. The reactions involved are as follows:

$$H_3PO_4 + 3C_9H_7N + 12Na_2MoO_4 + 24HNO_3 = (C_9H_7N)_3H_3[P(Mo_3O_{10})_4]\cdot H_2O + 11H_2O + 24NaNO_3$$

$$(C_9H_7N)_3H_3[P(Mo_3O_{10})_4]\cdot H_2O + 26OH^- = HPO_4^{2-} + 12MoSO_4^{2-} + 3C_9H_7N + 15H_2O$$

Overdose: $OH^- + H^+ = H_2O$

Materials

Equipment: basic burette(50 mL), electric stove, volumetric flask (250 mL), pipette (25 mL), graduated cylinder (50 mL), washing ear ball, filter paper, beaker, suction devices, analytical balance.

Reagents: NaOH solution (4 g/L), NaOH standard solution (0.5 mol/L), HCl standard solution (0.25 mol/L), thymol bromophenol blue-phenolphthalein indicator, high-content quinine reagent, hydrochloric acid (1 : 1), nitric acid (1 : 1), phosphate fertilizer sample.

Procedures

(1) Sample decomposition.

Weigh 1 g of sample in a 250 mL beaker, and then moisten the sample with a small amount of water (about 5 mL). Add 20～25 mL of hydrochloric acid and 7～9 mL of nitric acid, mix well and cover with a watch glass. Slowly heat it on an electric stove for 30 min (water can be supplied during this process to prevent from drying out). Remove the beaker and add 10 mL of hydrochloric acid (1 : 1) after cooling, then filter under reduced pressure, wash the beaker and filter paper 5 times with water (about 10 mL each time), transfer the filtrate to a 250 mL volumetric flask, dilute to mark line after cooling.

(2) Precipitation.

Transfer 25.00 mL of sample solution to 300 mL beaker, add 10 mL of nitric acid (1 : 1) solution, dilute to 100 mL with water, cover with the watch glass, pre-heat until nearly boiling. Add 50 mL of high-content quinine reagent, keep faint boiling and stirring for 1 min (the precipitate of phosphomolybdic acid quinoline is formed during this process), and then cool to room temperature.

(3) Filtration and washing.

The mixture obtained above was filtered under reduced pressure, and the supernatant was filtered firstly, and then the precipitate was washed with distilled water (about 25 mL) for 3～4 times by using a pouring method. Finally, the precipitate was transferred to a filter and washed 3～5 times with distilled water.

(4) Dissolution of the precipitate and titration.

Transfer the precipitate together with the filter paper into the original beaker. Add the NaOH standard solution dropwise with a burette, dissolve the yellow solid stuck in the Büchner funnel into the beaker, wait until the yellow disappears, rinse the funnel with water and collecting the washing in the beaker. Crush the filter paper, continue to add NaOH standard solution and stir well until the yellow precipitate is completely dissolved. Add 100 mL of fresh boiled water with stirring, and then add 1 mL of thymol bromophenol blue-phenolphthalein indicator, titrate with HCl standard solution until the color of the solution changes from purple (through the gray-blue) to yellow (or colorless) as the endpoint.

(5) Blank experiment.

Carry out the blank experiment according to the above steps.

Data Recording and Processing

According to the recorded data, calculate the percentage content of phosphorus pentoxide.

$$w_{P_2O_5}=\frac{1}{52}\times\frac{c_{NaOH}(V_1-V_3)-c_{HCl}(V_2-V_4)}{m\dfrac{25.00}{250.0}\times 1000}\times M_{P_2O_5}$$

Where, c_{NaOH} is the concentration of NaOH standard solution, mol/L; c_{HCl} is the concentration of HCl standard solution, mol/L; V_1 is the consumed volume of NaOH standard solution, mL; V_2 is the consumed volume of HCl standard solution, mL; V_3 is the consumed volume of NaOH standard solution in blank experiment, mL; V_4 is the consumed volume of HCl standard solution in blank experiment, mL; m is the mass of the phosphate fertilizer sample; 25.00 is the volume of the transferred test solution, mL; 250.0 is the total volume of the prepared test solution, mL; $M_{P_2O_5}$ is the molar mass of phosphorus pentoxide, g/mol; 1/52 is the molar ratio of the phosphorus pentoxide to sodium hydroxide.

Notes

Phosphomolybdate heteropoly acid is stable under acidic conditions, and it will decompose into simple acid ions under alkaline condition. Acidity, temperature, and concentration of coordination acid anhydride will seriously affect the composition of heteropoly acids. Therefore, the precipitation conditions must be strictly controlled. In theory, the precipitation is favorable for the high acidity. However, it is difficult to precipitate under higher acidity. The reaction is not complete if the acidity is too low, resulting in measurement errors.

Questions

(1) Describe the principle of phosphomolybdic acid quinoline volumetric method for the determination of phosphorus in phosphate fertilizer.

(2) How to control the measurement conditions for the determination of phosphorus in phosphate fertilizer by using phosphomolybdic acid quinoline volumetric method?

Exp.9 Determination of Total Acidity in Edible Vinegar

Objectives

(1) To understand the change of pH in the titration of weak acid with strong base and the selection of indicator.

(2) To learn the method for the determination of total acidity in edible vinegar.

Principles

The main acidic substance of edible vinegar is acetic acid (HAc), it also contains a small amount of other weak acids such as lactic acid, succinic acid, malic acid, amino acids, etc. The dissociation constant (K_a) of HAc is 1.8×10^{-5}, and it could be titrated by NaOH standard solution. The solution pH is about 8.7 at the stoichiometric point, so phenolphthalein can be employed as indicator. The color of the solution changes from colorless to slight reddish at the endpoint. HAc and other acids presented in the edible vinegar can react with NaOH during the titration process, therefore, the total acidity obtained by titration is expressed by ρ_{HAc} (g/L).

Materials

Equipment: basic burette (50 mL), pipette (25 mL), volumetric flask (250 mL), Erlenmeyer flask (250 mL).

Reagents: NaOH standard solution (0.1 mol/L), potassium hydrogen phthalate (KHP, primary standard), phenolphthalein indicator (2 g/L, in ethanol), edible vinegar.

Procedures

(1) Standardization of NaOH solution (0.1 mol/L).

See details in Exp. 3.

(2) Determination of total acidity in edible vinegar.

Accurately transfer 25.00 mL of edible vinegar in a 250 mL volumetric flask, dilute to the mark with freshly boiled and cooled distilled water, and shake well. Pipette 25.00 mL of the above solution to a 250 mL Erlenmeyer flask and add 2~3 drops of phenolphthalein indicator. Then, titrate with NaOH standard solution (0.1 mol/L) until the solution is slight reddish and kept for 30 s. Record the consumed volume of NaOH standard solution in Table 3.8. Repeat the titration 3 times.

Table 3.8 Determination of total acidity in edible vinegar

	1	2	3
$V_{vinegar}$/mL			
$V_{diluted}$/mL			
V_{NaOH} (final)/mL			
V_{NaOH} (initial)/mL			
V_{NaOH}/mL			
ρ_{HAc}/(g/L)			
$\bar{\rho}_{HAc}$/(g/L)			
$\lvert d_i \rvert$			
$\bar{d}_r$/%			

Data Recording and Processing

(1) Write the calculation formula, calculate the concentration of NaOH standard solution, as well as the relative standard deviation.

(2) According to the concentration and consumed volume of NaOH standard solution, calculate the total acidity of edible vinegar, the average deviation and relative average deviation ($\bar{d}_r \leqslant 0.3\%$).

Questions

Why the distilled water with removing carbon dioxide was used when measuring the content of acetic acid?

Exp.10 Determination of the Molar Mass of Organic Acids

Objectives

(1) To master the method for determining the molar mass of organic acids.

(2) To be familiar with the preparation and standardization of NaOH standard solution.

Principles

Most organic acids are weak acids, such as oxalic acid (pK_{a1}=1.23, pK_{a2}=4.19), tartaric acid (pK_{a1}=2.85, pK_{a2}=4.34) and citric acid (pK_{a1}=3.15, pK_{a2}=4.77, pK_{a3}=6.39), etc. Generally, these acids are soluble in water. If the concentration of acid is up to 0.1 mol/L and the $cK_a \geqslant 10^{-8}$, it can be titrated by NaOH standard solution with phenolphthalein as indicator. The color of the solution is slight reddish at the endpoint. The molar mass of the organic acid can be obtained based on the concentration and the consumed volume of NaOH standard solution, and the mass of the organic acid weighed. If the organic acid is a polyprotic acid, the stoichiometric relationship between the reaction of polyprotic acid and NaOH should be determined based on the discriminant formula of whether each proton can be accurately titrated ($c_{ai}\, K_{ai} \geqslant 10^{-8}$), and whether the two grades of acid can be fractionally titrated ($c_{ai}\, K_{ai} / c_{ai+1}\, K_{ai+1} \geqslant 10^{5}$), and then calculate the molar mass of organic acid.

Materials

Equipment: basic burette (50 mL), pipette (25 mL), volumetric flask (250 mL), Erlenmeyer flask (250 mL), beaker, analytical balance.

Reagents: NaOH standard solution (0.1 mol/L), potassium hydrogen phthalate (KHP, primary standard), phenolphthalein indicator (2 g/L, in ethanol), organic acid (such as oxalic acid, tartaric acid, citric acid, acetylsalicylic acid, etc.).

Procedures

(1) Preparation and standardization of NaOH solution(0.1 mol/L).

See details in Exp. 3.

(2) Determination of the molar mass of organic acids.

Accurately weigh the organic acid sample in a beaker, dissolve with distilled water, quantitatively transfer to a 250 mL volumetric flask, dilute to the mark with distilled water, and shake well. Pipette 25.00 mL of sample solution to a 250 mL Erlenmeyer flask, add 2~3 drops of phenolphthalein indicator, then titrate with NaOH standard solution until the color of the solution changes from colorless to pink and does not fade within 30 s, indicating the endpoint. Repeat the titration 3 times. Record the consumed volume of NaOH standard solution in Table 3.9.

Data Recording and Processing

(1) Write the calculation formula and calculate the concentration of NaOH standard solution, as well as relative average deviation.

(2) According to the concentration and the consumed volume of NaOH standard solution, calculate the molar mass of organic acid, as well as the average deviation and relative average deviation.

Table 3.9 Determination of molar mass of organic acid

	1	2	3
$m_{\text{organic acid}}$/g			
V_{sample} /mL			
V_{NaOH}(final)/mL			
V_{NaOH}(initial)/mL			
V_{NaOH}/mL			
$M_{\text{organic acid}}$/(g/mol)			
$\bar{M}_{\text{organic acid}}$/(g/mol)			
$\lvert d_i \rvert$			
$\bar{d}_r$ /%			

Questions

(1) If the NaOH standard solution absorbs carbon dioxide in the air during storage, what is the difference between using methyl orange and phenolphthalein as indicators while titrating the same HCl solution with this standard solution? Why?

(2) Can polyprotic acids such as oxalic acid, citric acid and tartaric acid be titrated with NaOH standard solution?

3.3 Complexometric Titration Experiments

Exp.11 Preparation and Standardization of EDTA Standard Solution and Determination of Water Hardness

Objectives

(1) To master the preparation and standardization of EDTA standard solution.

(2) To grasp the judgments of the endpoints with eriochrome black T and calcium indicator.

(3) To master the principle and method for the determination of water hardness by complexometric titration.

Principles

The four carboxyl oxygen atoms and two amino nitrogen atoms in the ethylenediamine tetraacetic acid (known as EDTA, or EDTA acid, simplified as H_4Y) molecule can coordinate with metal ion. EDTA is the most widely used chelating agent in analytical chemistry. The standard solution of EDTA is commonly prepared by dissolving disodium dihydrogen ethylenediamine tetraacetate ($Na_2H_2Y \cdot 2H_2O$, also simplified as EDTA, or EDTA disodium salt) in water, because EDTA acid is hardly soluble in water. It is difficult to get the pure form of EDTA. Therefore, the standard solution of EDTA should be prepared indirectly.

The total hardness of water refers to the total amount of Ca^{2+} and Mg^{2+} in water, and is classified into calcium hardness (hardness caused by Ca^{2+}) and magnesium hardness (hardness caused by Mg^{2+}). For industrial water, the hardness of water is a major factor to form pot scale and affect product quality. Therefore, the determination of hardness of water can provide a basis for the detection of the water quality and water treatment. The hardness of water is usually expressed by converting Ca and Mg into the total amount of CaO, and the unit is mg/L.

For determination of water hardness, the concentration of EDTA solution is standardized with $CaCO_3$ as the primary standard and calcium indicator (In) as the indicator at pH 12～13 adjusted with NaOH solution. The color of the solution turns from purple-red to blue indicating the endpoint. The relevant reactions are as follows:

Before the titration $\quad Ca + In\ (blue) = CaIn\ (purple\text{-}red)$

In the titration $\quad Ca + Y = CaY$

At the endpoint $\quad CaIn\ (purple\text{-}red) + Y = CaY + In\ (blue)$

The total hardness of water (total concentration of Ca^{2+} and Mg^{2+}) is generally determined by using EDTA standard solution and eriochrome black T (EBT) as the indicator at pH ≈ 10 with NH_3-NH_4Cl buffer solution. The interferences of Fe^{3+}, Al^{3+}, Cu^{2+}, Pb^{2+} and Zn^{2+} can be masked with triethanolamine.

To separately determine the hardness of Ca and Mg, the sample pH is adjusted to 12～13 with

NaOH solution to form $Mg(OH)_2$, then the concentration of Ca^{2+} is determined by titration using EDTA standard solution and calcium indicator, and the color of the solution changes from purple-red to blue indicating the endpoint. Finally, the concentration of Mg^{2+} is determined by the difference between total hardness and Ca hardness.

Materials

Equipment: basic burette (50 mL), beaker (100 mL, 500 mL), volumetric flask (250 mL), Erlenmeyer flask (250 mL), pipette (25 mL), analytical balance.

Reagents: disodium dihydrogen ethylenediamine tetraacetate ($Na_2H_2Y \cdot 2H_2O$, AR), $CaCO_3$ (GR), eriochrome black T (EBT) indicator, calcium indicator, NH_3-NH_4Cl buffer solution (pH $\approx$ 10), HCl solution (6 mol/L), NaOH solution (1 mol/L), triethanolamine solution (1 ∶ 2).

Procedures

(1) Preparation of EDTA standard solution (0.02 mol/L).

Weigh 4.0 g of $Na_2H_2Y \cdot 2H_2O$, then dissolve it with 200 mL of water in a 500 mL beaker under heating. After cooled, the solution is transferred to reagent bottle (if for reserving, stored in a polyethylene bottle), and diluted to 500 mL, mixed well and labeled.

(2) Preparation of Ca^{2+} standard solution (0.02 mol/L).

Accurately weigh 0.50 g～0.55 g of $CaCO_3$ (primary standard), put it in a 100 mL beaker, then add a few drops of water to wet, cover the watch glass, slowly add 5 mL of HCl solution (6 mol/L) to dissolve and then quantitative transfer to a 250 mL volumetric flask. Finally, dilute to the mark with distilled water and shake well.

(3) Standardization of EDTA standard solution.

Transfer 25.00 mL of Ca^{2+} standard solution to a 250 mL Erlenmeyer flask, add 5 mL of NaOH solution (1 mol/L) and a little amount of calcium indicator, then mixed and titrated with EDTA standard solution until the color of the solution turns from purple-red to blue. Record the consumed volume of EDTA standard solution (V_0) in Table 3.10 and repeat the titration 3 times.

(4) Determination of total hardness.

Transfer 25.00 mL of water sample to a 250 mL Erlenmeyer flask, add 5 mL of buffer solution (NH_3-NH_4Cl) and 3～4 drops of EBT indicator, then mixed well and titrate with EDTA standard solution until the color of the solution turns from purple-red to blue. Record the consumed volume of EDTA standard solution (V_1) in Table 3.11 and repeat the titration 3 times.

(5) Determination of the Ca hardness.

Transfer 25.00 mL of water sample to a 250 mL Erlenmeyer flask, add 5 mL of NaOH solution (1 mol/L) and a few of calcium indicator. Mix well and titrate with EDTA standard solution until the color of the solution turns from purple-red to blue. Record the consumed volume of EDTA standard solution (V_2) in Table 3.12 and repeat the titration 3 times.

Data Recording and Processing

(1) Calculate the concentration of EDTA standard solution, mean value and relative average deviation ($\bar{d}_r \leqslant 0.3\%$).

(2) Calculate the total hardness (mg/L) of water sample, Ca and Mg hardness, respectively, mean values and relative average deviations ($\bar{d}_r \leqslant 0.3\%$).

$$\rho_{tot} = \frac{c_{EDTA} \times V_1 \times M_{CaO}}{V_{water}} \times 1000$$

$$\rho_{Ca} = \frac{c_{EDTA} \times V_2 \times M_{Ca}}{V_{water}} \times 1000$$

$$\rho_{Mg} = \frac{c_{EDTA} \times (V_1 - V_2) \times M_{Mg}}{V_{water}} \times 1000$$

Table 3.10 Standardization of EDTA standard solution

	1	2	3
m_{CaCO_3}/g			
$c_{Ca^{2+}}$/(mol/L)			
V_0 (final)/mL			
V_0 (initial)/mL			
V_0/mL			
c_{EDTA}/(mol/L)			
$\bar{c}_{EDTA}$/(mol/L)			
$\|d_i\|$			
$\bar{d}_r$ /%			

Table 3.11 Determination of total hardness

	1	2	3
$V_{water\ sample}$/mL			
V_1 (final)/mL			
V_1(initial)/mL			
V_1/mL			
ρ_{tot}/(mg/L)			
$\bar{\rho}_{tot}$/(mg/L)			
$\|d_i\|$			
$\bar{d}_r$ /%			

Table 3.12 Determination of calcium hardness and magnesium hardness

	1	2	3
$V_{\text{water sample}}$/mL			
V_2(final)/mL			
V_2(initial)/mL			
V_2/mL			
ρ_{Ca}/(mg/L)			
$\bar{\rho}_{Ca}$ /(mg/L)			
$\lvert d_i \rvert$			
$\bar{d}_r$/%			
ρ_{Mg}/(mg/L)			
$\bar{\rho}_{Mg}$ /(mg/L)			
$\lvert d_i \rvert$			
$\bar{d}_r$/%			

Questions

(1) What is the function of the buffer solution in complexometric titration?

(2) What range of the pH should be controlled in the titration using EBT as the indicator? How to perform?

(3) Why should the titration be slow in this experiment?

Exp.12 Determination of Calcium and Magnesium in Industrial Raw Materials

Objectives

(1) To master the principle and method of the complexometric titration for determining calcium and magnesium in limestone.

(2) To learn the selection and application of the indicator in complexometric titration.

Principles

Limestone is an important industrial material. It mainly consists of $CaCO_3$, a certain amount of $MgCO_3$ and a few impurities such as Al, Fe, Si etc. Among them, the concentrations of calcium and magnesium can be determined by the complexometric titration.

The sample is dissolved by acid with Mg^{2+} and Ca^{2+} coexist in solution. The interference of Fe^{3+} and Al^{3+}can be prevented by adding triethanolamine. After the pH of the sample is adjusted to exceed 12, Mg^{2+} is transformed to the precipitate of $Mg(OH)_2$. Then, the concentration of calcium can be determined by titration using the EDTA standard solution with the calcium indicator (In). At the endpoint, the color of the solution turns from purple-red to blue. The concentration of calcium is represented as the mass fraction of CaO. The reaction principles are as below:

Before the titration $\quad Ca + In\ (bule) = CaIn\ (purple\text{-}red)$

In the titration $$Ca + Y \xlongequal{} CaY$$

At the endpoint $$CaIn\ (purple\text{-}red) + Y \xlongequal{} CaY + In\ (blue)$$

Take another sample and add the triethanolamine to mask the interfering ions (Fe^{3+}, Al^{3+}, etc.). After the pH of the sample is adjusted to about 10, the total concentration of calcium and magnesium can be determined by titration using the EDTA standard solution. With the eriochrome black T (EBT) as the indicator, the color of the solution turns from purple-red (via purple blue) to blue at the endpoint. The concentration of magnesium, which is represented as the mass fraction of MgO, can be calculated by subtracting the concentration of calcium from the total concentration. The reaction principles are as below:

Before the titration $$Mg + EBT\ (blue) \xlongequal{} Mg\text{-}EBT\ (purple\text{-}red)$$

In the titration $$Ca\ (Mg) + Y \xlongequal{} CaY\ (MgY)$$

At the endpoint $$Mg\text{-}EBT\ (purple\text{-}red) + Y \xlongequal{} MgY + EBT\ (blue)$$

Materials

Equipment: acid burette (50 mL), Erlenmeyer flask (250 mL), piette (25 mL), volumetric flask (250 mL), graduated cylinder (10 mL, 25 mL), beaker (100 mL), analytical balance.

Reagents: EDTA standard solution (0.02 mol/L), NaOH solution (10%), HCl solution (1 : 1), triethanolamine solution (1 : 2), NH_3-NH_4Cl buffer solution (pH ≈ 10), calcium indicator, eriochrome black T, limestone sample.

Procedures

(1) Preparation and standardization of EDTA standard solution (0.02 mol/L).

See details in Exp. 11.

(2) Preparation of test solution.

Accurately weigh 0.25 g～0.30 g of sample to a 100 mL beaker. Wet it with a few water, cover with the watch glass, and then slowly add 4～6 mL of the HCl solution (1 : 1). When the reaction is severe, stop for a moment, and turn the beaker slightly until the complete solvation. Use a small amount of water to clean the convex surface of the watch glass and the inner wall of the beaker, and make all the washing liquid flow into the beaker. Transfer the solution to a 250 mL volumetric flask, dilute to the mark with distilled water and shake well.

(3) Determination of Ca.

Pipette 25.00 mL of the test solution to a 250 mL Erlenmeyer flask, add 20 mL of water and 5 mL of the triethanolamine solution and swirl. Add 10 mL of NaOH solution (10%) to make the pH of the solution in the range of 12～14. Then, add a little of calcium indicator and mix well. Titrate with EDTA standard solution until the color of the solution turns from purple-red to blue (slow down and shake more near the endpoint because the reaction speed is slow). Record the consumed volume of the EDTA standard solution (V_1) in Table 3.13 and repeat the titration 3 times.

(4) Determination of the total amount of Ca and Mg.

Accurately pipette 25.00 mL of the test solution to a 250 mL Erlenmeyer flask. Add 20 mL of distilled water, 5 mL of triethanolamine solution, 10 mL of NH_3-NH_4Cl buffer solution and 3~5 drops of EBT indicator. Mix well and titrate with EDTA standard solution until the color of the solution turns from purple-red to blue. Record the consumed volume of EDTA standard solution (V_2) in Table 3.13 and repeat the titration 3 times.

Data Recording and Processing

Calculate the mass fractions of CaO and MgO in the limestone sample, mean values and relative average deviations ($\bar{d}_r \leqslant 0.3\%$).

$$w_{CaO} = \frac{c_{EDTA} \times V_1 \times M_{CaO}}{25.00 \times m_{sample}} \times 250 \times 10^3 \times 100\%$$

$$w_{MgO} = \frac{c_{EDTA} \times (V_2 - V_1) \times M_{MgO}}{25.00 \times m_{sample}} \times 250 \times 10^3 \times 100\%$$

Table 3.13　Determination of calcium and magnesium content

	1	2	3
m_{sample}/g			
V_1(final)/mL			
V_1(initial)/mL			
V_1/mL			
w_{CaO}/%			
$\bar{w}_{CaO}$/%			
$\lvert d_i \rvert$			
$\bar{d}_r$/%			
V_2(final)/mL			
V_2(initial)/mL			
V_2/mL			
w_{MgO}/%			
$\bar{w}_{MgO}$/%			
$\lvert d_i \rvert$			
$\bar{d}_r$/%			

Questions

(1) To explain the principle of the complexometric titration for determining calcium and magnesium in limestone.

(2) What is the function of the triethanolamine solution in the determination of Ca^{2+} and Mg^{2+}? Is it possible to add triethanolamine after adding buffer solution? Why?

Exp.13 Determination of Pb^{2+} and Bi^{3+} in Mixture of Pb^{2+} and Bi^{3+}

Objectives

(1) To learn the principle to increase the selectivity of the complexometric titration by controlling acidity.

(2) To master the procedure of continuous titration of various metal ions using EDTA.

(3) To be familiar with the application of xylenol orange (XO) and the determination of endpoint.

Principles

Both Bi^{3+} and Pb^{2+} can form stable complexes with EDTA, but their formation constants are quite different ($\lg K_{BiY}$ =27.94, $\lg K_{PbY}$ =18.04), which meets the stepwise titration conditions of mixed ions (when $c_M = c_N$, $\Delta pM = \pm 0.2$, if need $|E_t| \leqslant 0.1\%$, $\Delta \lg K$ must be more than 6). Thus, the concentration of Bi^{3+} and Pb^{2+} can be successively determined by continuous titration through controlling the acidity of the mixed solution with xylenol orange (XO) as the indicator.

After adjusting the solution pH ≈ 1, Bi^{3+} forms a stable purple-red complex with XO (Pb^{2+} does not interfere under this condition). Then, titrate with EDTA standard solution until the color of the solution turns from purple-red to red, orange, finally to bright yellow, which is the first endpoint of the titration. Then add hexamethylene tetramine solution to adjust the pH in the range of 5～6. At this time Pb^{2+} forms purple complex with XO. Pb^{2+} is titrated using EDTA standard solution until the color of the solution turns from purple-red to bright yellow, which is the second endpoint.

In order to carry out the standardization and measurement under the same conditions, the concentration of EDTA solution should be standardized using $ZnSO_4 \cdot 7H_2O$ with xylenol orange as the indicator. The titration was carried out in the HCl-$(CH_2)_6N_4$ buffer solution with pH of 5～6 and the color change of the solution at the endpoint is the same as above.

Materials

Equipment: acid burette (50 mL), Erlenmeyer flask (250 mL), pipette (25 mL), volumetric flask (250 mL), beaker (100 mL), analytical balance, etc.

Reagents: EDTA standard solution (0.02 mol/L), $ZnSO_4 \cdot 7H_2O$ (primary standard), HNO_3 solution (0.1 mol/L), HCl solution (1 : 5), hexamethylene tetramine solution (20%, AR), xylenol orange solution (0.2%), mixed solution of Bi^{3+} and Pb^{2+} ($c_{Bi^{3+}} \approx 0.01$ mol/L, $c_{Pb^{2+}} \approx 0.01$ mol/L, $c_{HNO_3} \approx 0.15$ mol/L).

Procedures

(1) Preparation of Zn standard solution (0.02 mol/L).

Accurately weigh 1.40～1.45 g of $ZnSO_4 \cdot 7H_2O$ primary standard in a 100 mL beaker, dissolve it with distilled water. Quantitatively transfer this solution to a 250 mL volumetric flask, dilute to the mark with distilled water and shake well.

(2) Preparation and standardization of EDTA standard solution (0.02 mol/L).

For the preparation of EDTA standard solution, see details in Exp. 11.

Accurately pipette 25.00 mL of Zn standard solution in a 250 mL Erlenmeyer flask. Add 2 mL of HCl solution (1 : 5), 2 drops of xylenol orange as indicator, add hexamethylene tetramine solution drop by drop until the color of the solution is stable purple-red. Then add another more 5 mL of hexamethylene tetramine solution, and shake well. Titrate the solution with EDTA standard solution until the color of the solution changes from magenta to bright yellow (when the endpoint is approached, drop slowly and shake intensely to avoid overdose). Record the consumed volume of EDTA (V_{EDTA}) in Table 3.14.. Repeat the titration 3 times.

(3) Continuous determination of Bi^{3+} and Pb^{2+}.

Accurately transfer 25.00 mL of the mixed solution of Bi^{3+} and Pb^{2+} to a 250 mL Erlenmeyer flask. Add 10 mL of HNO_3 solution (0.1 mol/L) and 2 drops of XO solution, and shake well. Titrate the solution with EDTA standard solution until the color of the solution changes from purple-red to bright yellow, which is the first endpoint. Record the consumed volume of EDTA standard solution (V_{Bi}) in Table 3.15. Because of the slow reaction between Bi^{3+} and EDTA, the titration speed should be slow near the endpoint, and oscillate the solution with force.

Supplement another one drop of XO solution, add hexamethylene tetramine solution until the solution is stable purple-red, and then add another more 5 mL. The pH of the solution should be 5~6. Continue to titrate with EDTA standard solution until the purple-red turns to bright yellow, which is the second endpoint. Record the consumed volume of EDTA standard solution V_{tot} ($V_{Pb} = V_{tot} - V_{Bi}$) in Table 3.15. Repeat the titration 3 times.

Data Recording and Processing

(1) Calculate the concentration of EDTA standard solution, mean value and relative average deviation ($\bar{d}_r \leqslant 0.2\%$).

(2) Calculate the concentrations of lead and bismuth in the solution (g/L), mean values and the relative average deviations ($\bar{d}_r \leqslant 0.2\%$).

Table 3.14　Standardization of EDTA standard solution

	1	2	3
$m_{ZnSO_4 \cdot 7H_2O}$ /g			
V_{EDTA}(final)/mL			
V_{EDTA}(initial)/mL			
V_{EDTA}/mL			
c_{EDTA}/(mol/L)			
$\bar{c}_{EDTA}$ /(mol/L)			
$\lvert d_i \rvert$			
$\bar{d}_r$/%			

Table 3.15 Determination of lead and bismuth content

	1	2	3
V_{Bi}(final)/mL			
V_{Bi}(initial)/mL			
V_{Bi}/mL			
V_{tot}/mL			
V_{Pb}/mL			
ρ_{Bi}/(g/L)			
$\bar{\rho}_{Bi}$/(g/L)			
$\lvert d_i \rvert$			
$\bar{d}_r$/%			
ρ_{Pb}/(g/L)			
$\bar{\rho}_{Pb}$/(g/L)			
$\lvert d_i \rvert$			
$\bar{d}_r$/%			

Notes

In titration of Bi^{3+}, if the acidity is too low, Bi^{3+} will hydrolyze to generate white turbidity, which will cause the endpoint to appear prematurely and produce reddening. At this point, the solution should be placed for a moment and then continue to titrate until the solution is stable and bright yellow, which is the endpoint.

Questions

(1) What range of the pH should be controlled in the titration of Pb^{2+} and Bi^{3+}? How to adjust the pH?

(2) Why is the hexamethylene tetramine instead of sodium acetate added to adjust the solution pH to 5～6 while titrating Pb^{2+}?

(3) What is the effect of adding HNO_3 before titrating Bi^{3+}?

Exp.14 Determination of Aluminum and Magnesium in Gastropine Tablet

Objectives

(1) To learn the pretreatment method for the determination of the components in pharmaceutical products.

(2) To master the principle and operation of back titration.

Principles

Gastropine, a commonly used stomach medicine, is mainly composed of aluminum hydroxide [$Al(OH)_3$], magnesium trisilicate ($2MgO \cdot 3SiO_2 \cdot xH_2O$) and a small amount of belladonna extract and dextrin. The contents of aluminum and magnesium in Gastropine tablets can be determined by complexometric titration. The content of aluminum hydroxide can be obtained by the following procedures: firstly, dissolve the tablets with nitric acid, separate the insoluble matter, and then add the excess and known amount of EDTA solution, adjust the pH to 3～4, boil for a few minutes to make the Al^{3+} ions react with EDTA completely; then adjust the solution pH to 5～6 after cooling. The excess EDTA is back titrated with Zn^{2+} standard solution with xylenol orange as indicator.

Take another test solution, adjust its pH to 8～9, Al^{3+} will be separated by precipitation. Then Mg^{2+} in the filtrate can be titrated by EDTA with eriochrome black T as indicator at pH about 10.

Materials

Equipment: acid burette (50 mL), beaker (100 mL, 500 mL), volumetric flask (250 mL), Erlenmeyer flask (250 mL), piette (10 mL, 25 mL), graduated cylinder (10 mL, 50 mL), analytical balance, electric furnace, mortar, etc.

Reagents: EDTA standard solution (0.02 mol/L), Zn standard solution (0.02 mol/L , see details in Exp.13), hexamethylene tetramine solution (20%), NH_3-NH_4Cl buffer solution (pH≈10), triethanolamine solution (1∶2), HCl solution (1∶1), NaOH solution (1 mol/L), $NH_3 \cdot H_2O$ (1∶1), methyl red indicator (0.2%, in ethanol), EBT indicator, xylenol orange indicator(0.2%).

Procedures

(1) Preparation and standardization of EDTA standard solution (0.02 mol/L).

See details in Exp. 13.

(2) Sample preparation.

Take 10 tablets of Gastropine and ground, weigh 2 g of powder in a 250 mL beaker, add 20 mL of HCl solution (1∶1), dilute to 100 mL with distilled water and boil. After cooling, filter and wash the precipitate with distilled water. Collect filtrate and washing liquid in a 250 mL volumetric flask, dilute to the mark with distilled water and shake well.

(3) Determination of aluminum.

Accurately transfer 10.00 mL of the test solution in a 250 mL Erlenmeyer flask, dilute to 25 mL with distilled water. Add $NH_3 \cdot H_2O$ solution (1∶1) dropwise until turbidity appears, and then add HCl solution (1∶1) to dissolve the precipitation. Add 25.00 mL of EDTA standard solution and boil for 1 min. After cooling, add 10 mL of hexamethylene tetramine solution (20%) and 2～3 drops of xylenol orange indicator. Titrate with Zn standard solution until the color of the solution turns from yellow to red as the endpoint. Repeat the titration 3 times. Record the consumed volume of Zn standard solution (V_1) in Table 3.16.

(4) Determination of magnesium.

Transfer 25.00 mL of the test solution in a small beaker, add $NH_3 \cdot H_2O$ solution (1∶1) dropwise until the precipitation just appeared, then add HCl solution (1∶1) until the precipitate is just dissolved. Add 2 g of NH_4Cl and hexamethylene tetramine solution until the precipitate appears, then supplement another more 15 mL. The test solution is heated to 80℃ for 5 min, cooled, filtered, and the precipitate is washed several times with a small amount of distilled water.

The filtrate and washing solution are collected in a 250 mL Erlenmeyer flask. Then add 4 mL of triethanolamine (1 : 2), 10 mL of NH_3-NH_4Cl buffer solution (pH ≈ 10), 1 drop of methyl red indicator, 3～5 drops of EBT indicator, and titrate with EDTA standard solution until the color of the solution changes from dark red to blue-green as the endpoint. Titrate 3 times in parallel. Record the consumed volume of EDTA standard solution (V_2) in Table 3.16.

Data Recording and Processing

(1) Calculate the content of Al in the tablets [expressed as $Al(OH)_3$].

(2) Calculate the content of Mg in the tablets (expressed as MgO).

Table 3.16 Determination of aluminum and magnesium content

	1	2	3
V_1 (final)/mL			
V_1(initial)/mL			
V_1/mL			
$w_{Al(OH)_3}$ /%			
$\lvert d_i \rvert$			
$\bar{d}_r$/%			
V_2 (final)/mL			
V_2(initial)/mL			
V_2/mL			
w_{MgO} /%			
$\lvert d_i \rvert$			
$\bar{d}_r$/%			

Notes

(1) The content of aluminum and magnesium in each tablet of Gastropine may not be equal. To make the measurement results representative, several tablets of Gastropine are ground together and a small amount of powder is taken for analysis.

(2) One drop of methyl red can make the endpoint more sensitive when determining the content of Mg.

Questions

(1) Why do not use the direct titration method to determine Al^{3+}?

(2) What is the function of triethanolamine in the determination of Mg^{2+}?

Exp.15 Determination of EDTA in Detergents

Objectives

(1) To master the preparation and standardization of copper sulphate solution.

(2) To master the principle and operation for determination of EDTA in detergents.

Principles

EDTA is an important component in detergent. It masks metal ions, which can increase the activity of surfactants and the stability of foaming, thus improve the cleaning efficiency of detergents. After adjusting the pH of the test solution to 4～5 with hydrochloric acid, EDTA in the sample can be titrated by copper sulfate standard solution with the 1-(2-pyridylazo)-2-naphthol (PAN) as indicator. The PAN itself is yellow in this pH range, and its complexes with metal ions are red. Therefore, the color change of the solution at the endpoint is from yellow-green [CuY (blue) + PAN (yellow)] to purple [CuY (blue) + Cu-PAN (red)].

Materials

Equipment: micro burette (2 mL), volumetric flask (250 mL), Erlenmeyer flask (250 mL), pipette (25 mL), analytical balance (0.1 mg).

Reagents: EDTA standard solution (0.02 mol/L), 1-(2-pyridylazo)-2-naphthol (PAN) solution (1 g/L, in ethanol), HCl solution (5 mol/L), $CuSO_4 \cdot 5H_2O$ (s, AR), acetic acid buffer solution (pH ≈ 4.65, mixed by 0.4 mol/L acetic acid solution and 0.2 mol/L NaOH solution in equal volume), detergent sample.

Procedures

(1) Preparation and standardization of $CuSO_4$ standard solution (0.01 mol/L).

Weigh 2.5 g of copper sulfate ($CuSO_4 \cdot 5H_2O$), dissolve and dilute to 1000 mL with distilled water. The standardization of the prepared copper sulfate solution with 0.02 mol/L EDTA standard solution (see details in Exp.11) is as follows. Accurately pipette 10.00 mL of EDTA standard solution (0.02 mol/L) to a 250 mL Erlenmeyer flask, dilute with 50 mL distilled water, adjust pH to 4～5 with HCl solution. Add 5 mL of acetic acid buffer solution and heat the solution to 60℃, supplement 5～6 drops of PAN indicator. Titrate with $CuSO_4$ standard solution until the color of the solution changes from yellow to purple and keeps for 1 min. Record the consumed volume of $CuSO_4$ solution (V_1) in Table 3.17. Repeat the titration 3 times.

(2) Determination of EDTA in detergent.

Accurately weigh 5.0～5.5 g of sample in a 250 mL Erlenmeyer flask, dilute with water, adjust solution pH to 4～5 with HCl solution. Add 5 mL of acetic acid buffer solution and heat the solution to 60℃, supplement 5～6 drops of PAN indicator. Titrate with $CuSO_4$ standard solution by micro burette under stirring until the color of the solution changes from yellow to purple and keeps for 1 min. Record the consumed volume of $CuSO_4$ standard solution (V_2) in Table 3.18. Repeat the titration 3 times.

Data Recording and Processing

(1) Calculate the concentration of $CuSO_4$ standard solution, mean value and relative average deviation ($\bar{d}_r \leqslant 0.2\%$).

(2) Calculate the mass fraction of EDTA in detergent sample, mean value and relative average deviation ($\bar{d}_r \leqslant 0.2\%$).

Table 3.17 Standardization of copper sulfate standard solution

	1	2	3
V_1(final)/mL			
V_1(initial)/mL			
V_1/mL			
c_{CuSO_4}/(mol/L)			
$\bar{c}_{CuSO_4}$/(mol/L)			
$\|d_i\|$			
$\bar{d}_r$/%			

Table 3.18 Determination of EDTA content in sample

	1	2	3
V_2(final)/mL			
V_2(initial)/mL			
V_2/mL			
w_{EDTA}/%			
$\bar{w}_{EDTA}$/%			
$\|d_i\|$			
$\bar{d}_r$/%			

Notes

(1) Generally, the content of EDTA in detergent is low, thus it is necessary to titrate slowly with continuous stirring to avoid overdose.

(2) The pH of the sample must be strictly controlled, pH meter or precise pH test paper can be used.

Questions

(1) Describe the principle for the determination of EDTA in detergent.

(2) Is there any other method to determine EDTA? Give an example.

(3) What reaction will be involved between Cu^{2+} and EDTA at different pH?

Exp.16 Determination of Iron, Aluminum, Calcium and Magnesium in Cement Clinker

Objectives

(1) To consolidate the principle for determination of the coexisting components by controlling the acidity in the complexometric titration.

(2) To learn the method of eliminating interference components in complexometric titration.

Principles

The main chemical components of cement clinker are SiO_2 (18%～24%), Fe_2O_3 (2.0%～5.5%), Al_2O_3 (4.0%～9.5%), CaO (60%～67%) and MgO (<4.5%). While reacting with hydrochloric acid, they can form silicic acid and some soluble chlorides. Silicic acid is a kind of inorganic acid, which exists as sol state in aqueous solution. Its chemical formula is $SiO_2 \cdot H_2O$. After dealing with the concentrated acid and heating, most hydrosol of silicic acid will be dehydrated into the hydrogel and separated out. Therefore, the precipitation separation method can separate silicic acid from iron, aluminum, calcium and magnesium in cement.

Iron, aluminum, calcium, magnesium exist as Fe^{3+}, Al^{3+}, Ca^{2+}, Mg^{2+} in the filtrate after the SiO_2 is precipitated. All of these ions can form the stable complexes with EDTA ($\lg K_{AlY}$ = 16.3, $\lg K_{CaY}$ = 10.69 and $\lg K_{MgY}$ = 8.7), There are marked differences in the stability, thus they can be titrated by EDTA standard solution by controlling appropriate acidity.

(1) Determination of iron.

In the solution pH of 1.5～2.5 and temperature of 60～70℃, iron can be titrated with EDTA standard solution by using the sulfosulicylic acid or its sodium salt as the indicator. The color of the solution changes from purple-red to bright yellow at the endpoint.

Chromogenic reaction $Fe^{3+} + HIn^- = FeIn^+$ (purple-red) $+ H^+$

Titration reaction $Fe^{3+} + H_2Y^{2-} = FeY^- + 2H^+$

At the endpoint $FeIn^+$ (purple-red) $+ H_2Y^{2-} = FeY^-$(bright yellow) $+ HIn^- + H^+$

(2) Determination of aluminum.

The coordination reaction is very slow between Al^{3+} and EDTA, which is unfit for direct titration. Add excess EDTA standard solution and heat to boiling to finish the reaction between Al^{3+} and EDTA completely at pH ≈ 4.3. Then use copper salt standard solution to back titrate with the PAN as indicator. The solution is yellow (color of indicator) at the beginning, and changes from yellow to green (mixed color of PAN and CuY). At the endpoint, the solution is purple (mixed color of PAN and CuPAN).

Titration reaction $Al^{3+} + H_2Y^{2-}$ (excess) $= AlY^-$ (colorless) $+ 2H^+$

Back titrate of the excess EDTA $Cu^{2+} + H_2Y^{2-}$ (remain) $= CuY^{2-}$ (blue) $+ 2H^+$

At the endpoint Cu^{2+} + PAN (yellow) = Cu-PAN (red)

The quantity of the blue CuY in solution affects the sensitivity of color changing at the endpoint. So the excess amount of EDTA should be controlled. Generally, an excess of 10～15 mL of EDTA standard solution (0.01～0.015 mol/L) in 100 mL of the solution is suggested.

(3) Determination of calcium.

In a strong base solution with pH above 12, Mg^{2+} will be masked when it forms precipitation of $Mg(OH)_2$. Fe^{3+} and Al^{3+} can be masked by triethanolamine. Then calcium can be titrated with EDTA standard solution using calcein-methyl thymol blue-phenolphthalein (CMP) as the mixed indicator.

Calcein itself is orange at pH > 12. It presents green florescence while coordinating with Ca^{2+}, Sr^{2+}, Ba^{2+}, etc. At the endpoint, the solution is orange. But the residual fluorescence in the solution will affect the observation of endpoint. In this experiment, the methyl thymol blue and phenolphthalein in mixed indicator is employed to cover the residual florescence of calcein in the titration.

(4) Determination of magnesium.

The content of magnesium is usually determined by subtraction method using EDTA complexometric titration. Adjust the solution pH ≈ 10, titrate the total content of calcium and magnesium by EDTA, and then subtract the amount of calcium to obtain the content of magnesium.

While titrating the total content of calcium and magnesium, eriochrome black T and acid chrome blue K-naphthol green B (K-B) are usually used as the indicators. Eriochrome black T is easily masked by some heavy metal ions. In this experiment, K-B mixed indicator is selected for the titration of the total content of calcium and magnesium with EDTA. Naphthol green B itself has no color change during the titration process, but it can stand out the ending color change from red to blue. Fe^{3+} and Al^{3+} can be masked by triethanolamine and potassium sodium tartrate.

Materials

Equipment: beaker (100 mL, 1000 mL), graduated cylinder (100 mL), pipette (5 mL, 10 mL, 25 mL), volumetric flask (250 mL), Erlenmeyer flask (250 mL), acid burette (50 mL), analytical balance.

Reagents: EDTA standard solution (0.01 mol/L, see Exp. 11), $CuSO_4$ standard solution (0.01 mol/L, see Exp. 15), sulfosalicylic acid (10%), bromocresol green indicator (0.05%), PAN indicator (0.3%), K-B indicator, CMP indicator (mix 1 g of calcein, 1 g of methyl thymol blue, 0.2 g of phenolphthalein and 50 g of potassium nitrate dried at 105℃, grind and store in ground bottle), HAc-NaAc buffer solution (pH ≈ 4.3, dissolve 33.7 g of anhydrous sodium acetate in water, add 80 mL of glacial acetic acid, diluted with water to 1 L and shake well), seignette salt (10%, dissolve 10 g of seignette salt in 90 mL water), NH_3-NH_4Cl buffer (pH ≈ 10), HCl solution (6 mol/L), concentrated HCl, concentrated HNO_3, KOH solution (20%), ammonia solution (1 : 1), triethanolamine solution (1 : 2).

Procedures

Accurately weigh about 0.5 g of sample in 100 mL beaker, add 10 mL of concentrated HCl and 1 mL of concentrated HNO_3 along the beaker mouth with carefully stirring until the entire dark gray sample becomes pale yellow paste. Then cover it with the watch glass and heat the beaker on a hot plate (or boiling water bath) in the fume hood to dissolve the soluble salts. Filtrate and put the filtrate into a 250 mL volumetric flask. Scrub the glass rod and the beaker with hot hydrochloric acid (3 : 97) by a dropper and wash the residue for 3～4 times. Then wash the residue with hot water thoroughly (usually about 10 times, until no chloride ion). Put the filtrate and washing liquid together into a 250 mL volumetric flask, cool down to room temperature, dilute to mark with distilled water and shake well. The sample solution is prepared for later use.

(1) Determination of Fe_2O_3.

Accurately pipette 25.00 mL of sample solution in 250 mL Erlenmeyer flask, add 50 mL of water, 2 drops of 0.05% bromocresol green indicator (it is yellow at pH < 3.8, green at pH > 5.4). Add ammonia (1 : 1) dropwise until it is green. Then add HCl solution (6 mol/L) to make the solution become yellow. Supplement another 3 drops of HCl solution. The pH of the solution is in the range of 1.8～2.0. Heat the solution to 60～70℃, add 10 drops of sulfosalicylic acid indicator. Titrate with EDTA standard solution slowly until the color of the solution changes from purple-red to bright yellow (the temperature of the solution at the endpoint is not lower than 60℃, and then reserve the solution for determining the content of Al_2O_3). Record the consumed volume of EDTA standard solution (V_{EDTA}) in Table 3.19.

(2) Determination of Al_2O_3.

After the titration of iron, add 25.00 mL of EDTA standard solution (0.01 mol/L) accurately and 15 mL of HAc-NaAc buffer solution (pH ≈ 4.3). Boil it for 1～2 min, add 4～5 drops of PAN indicator after cooling down. Titrate with $CuSO_4$ standard solution (0.01 mol/L) until the solution is bright purple. Record the consumed volume of $CuSO_4$ standard solution (V_{CuSO_4}) in Table 3.20.

(3) Determination of CaO.

Accurately transfer 10.00 mL of sample solution in a 250 mL Erlenmeyer flask, add 50 mL of distilled water, 5 mL of triethanolamine solution (1 : 2) and a little CMP indicator. Then add 20% KOH solution while stirring until green fluorescence appears. Supplement another 5～8 mL of 20% KOH (solution pH > 13). Titrate with EDTA standard solution until the solution is red, indicating the endpoint. Record the consumed volume of EDTA standard solution (V_1) in Table 3.21.

(4) Determination of MgO.

Accurately transfer 10.00 mL of sample solution in a 250 mL Erlenmeyer flask, add 50 mL of distilled water, 1 mL of seignette salt and 5 mL of triethanolamine solution (1 : 2), shake for 1 min. Then add 15 mL of NH_3-NH_4Cl buffer (pH ≈ 10) and a few of K-B indicator. Titrate with EDTA standard solution (0.01 mol/L) until the color of the solution changes from purple-red to blue. Record the consumed volume of EDTA standard solution (V_2) in Table 3.22.

Data Recording and Processing

Calculate the content of iron, aluminum, calcium and magnesium in the cement sample, respectively.

$$w_{Fe_2O_3}=\frac{c_{EDTA}\times V_{EDTA}\times M_{Fe_2O_3}\times 250}{2m_s\times 1000\times 25.00}\times 100\%$$

$$w_{Al_2O_3}=\frac{(c_{EDTA}\times 25.00-c_{CuSO_4}\times V_{CuSO_4})\times M_{Al_2O_3}\times 250}{2m_s\times 1000\times 25.00}\times 100\%$$

$$w_{CaO}=\frac{c_{EDTA}\times V_1\times M_{CaO}\times 250}{m_s\times 1000\times 10.00}\times 100\%$$

$$w_{MgO}=\frac{c_{EDTA}\times (V_2-V_1)\times M_{MgO}\times 250}{m_s\times 1000\times 10.00}\times 100\%$$

Table 3.19 Determination of Fe_2O_3

	1	2	3
V_{EDTA}(final)/mL			
V_{EDTA}(initial)/mL			
V_{EDTA}/mL			
$w_{Fe_2O_3}$/%			
$\overline{w}_{Fe_2O_3}$/%			
$\|d_i\|$			
$\overline{d}_r$ /%			

Table 3.20 Determination of Al_2O_3

	1	2	3
V_{CuSO_4} (final)/mL			
V_{CuSO_4} (initial)/mL			
V_{CuSO_4} /mL			
$w_{Al_2O_3}$ /%			
$\overline{w}_{Al_2O_3}$ /%			
$\|d_i\|$			
$\overline{d}_r$ /%			

Table 3.21 Determination of CaO

	1	2	3
V_1(final)/mL			
V_1(initial)/mL			
V_1/mL			
w_{CaO}/%			
$\overline{w}_{CaO}$/%			
$\|d_i\|$			
$\overline{d}_r$ /%			

Table 3.22 Determination of MgO

	1	2	3
V_2(final)/mL			
V_2(initial)/mL			
V_2/mL			
w_{MgO}/%			
$\overline{w}_{MgO}$/%			
$\|d_i\|$			
$\overline{d}_r$ /%			

Questions

(1) How to eliminate the interference of Al^{3+}, Ca^{2+} and Mg^{2+} while titrating Fe^{3+}?

(2) What is the effect on the measurement of Al^{3+}, if the determination of Fe^{3+} is not accurate?

(3) Why do we use the back titration method to titrate Al^{3+} with EDTA?

(4) What is the purpose of adding triethanolamine in the determination of Ca^{2+} and Mg^{2+}? And why should triethanolamine be added before KOH in the determination of Ca^{2+}?

Exp.17 Determination of Aluminum in Industrial Aluminum Sulfate

Objectives

(1) To learn the basic principle of back titration.

(2) To master the principle and method of replacement titration for determining aluminum.

Principles

Al^{3+} is easy to form polynuclear hydroxyl complexes (such as $[Al_2(H_2O)_6(OH)_3]^{3+}$, $[Al_2(H_2O)\ (OH)_6]^{3+}$, etc.). The speed of complexation between Al^{3+} and EDTA is slow, therefore excess EDTA and boiling are needed to make sure the complexation is complete. It is better to use back titration or displacement titration rather than direct titration.

For back titration, an excess and known amount of EDTA standard solution is used and the pH is adjusted to about 3.5. The solution is boiled for a few minutes to make sure that Al^{3+} and EDTA react completely. Then the pH is adjusted to 5～6 and excess EDTA is back titrated with $CuSO_4$ standard solution.

For displacement titration, excess NH_4F is added to the solution, and the solution is heated to release the same amount of EDTA, because F^- and Al^{3+} can form a more stable complex. The released EDTA is titrated with $CuSO_4$ standard solution.

If the concentration of EDTA standard solution is c_{EDTA} (mol/L), the volume of $CuSO_4$ standard solution consumed in the standardization is V_1 (mL), the volume of $CuSO_4$ standard solution consumed in the titration of sample is V_2 (mL), the mass of sample is m (g), then the content of aluminum in the sample can be calculated by the formula below.

$$w_{Al}=\frac{26.98\times c_{EDTA}\times V_2\times 25.00\times 250}{V_1\times m\times 25.00}\times 100\%$$

Materials

Equipment: acid burette (25 mL), Erlenmeyer flask (250 mL), pipette (25 mL), beaker (100 mL), analytical balance.

Reagents: EDTA standard solution (0.02 mol/L), $CuSO_4\cdot 5H_2O$, PAN indicator (0.1%, 0.1 g of PAN dissolved in 100 mL of anhydrous ethanol), thymol blue indicator (0.1%, 0.1 g of thymol blue dissolved in 100 mL of 20% ethanol), HCl (1 : 1), H_2SO_4(1 : 1), $NH_3\cdot H_2O$ (1 : 1), NH_4F, hexamethylene tetramine buffer solution (20%), industrial aluminum sulfate.

Procedures

(1) Preparation and standardization of EDTA standard solution (0.02 mol/L).

See details in Exp. 11.

(2) Preparation and standardization of $CuSO_4$ standard solution (0.02 mol/L).

Weigh 2.6 g of $CuSO_4\cdot 5H_2O$, add 2～3 drops of H_2SO_4 solution (1 : 1), dissolve with 500 mL of water and mix well. Accurately transfer 25.00 mL of EDTA standard solution to a 250 mL Erlenmeyer flask. Add 10 mL of 20% hexamethylene tetramine buffer solution and heat to 80～90℃. Then add 2～3 mL of 0.1% PAN indicator and titrate with $CuSO_4$ standard solution until the solution is purple-red. Record the consumed volume of $CuSO_4$ standard solution (V_1) in Table 3.23. Repeat the titration 3 times.

(3) Determination of aluminum.

Accurately weigh 1.3 g of industrial aluminum sulfate in a 100 mL beaker, add 10 mL of HCl solution (1 : 1) and 50 mL of distilled water to dissolve, transfer to a 250 mL volumetric flask, dilute to the mark with distilled water and mix well.

Transfer 25.00 mL of the test solution to a 250 mL Erlenmeyer flask. Add 30 mL of EDTA standard solution, 5 drops of thymol blue indicator and several drops of $NH_3\cdot H_2O$ (1 : 1) until the solution is yellow (pH ≈ 3). Boil the solution for 2 min, add 10 mL of the hexamethylene tetramine buffer solution (20%) and 2～3 mL of PAN indicator. Then titrate with $CuSO_4$ standard solution (while the solution is hot) until the solution is purple-red. At this point, it is not necessary to record the consumed volume of $CuSO_4$ standard solution. Add 1～2 g of NH_4F and boil for 2 min (add another 8 drops of PAN indicator if necessary). Then titrate with $CuSO_4$ standard solution until the solution is purple-red again. Record the consumed volume of $CuSO_4$ standard solution (V_2) in Table 3.24. Repeat the titration 3 times.

Data Recording and Processing

Calculate the concentrations of EDTA and $CuSO_4$ standard solution, and the mass fraction of aluminum in the sample.

Table 3.23 Standardization of copper sulfate solution

	1	2	3
V_1(final)/mL			
V_1(initial)/mL			
V_1/mL			
c_{CuSO_4} /(mol/L)			
$\bar{c}_{CuSO_4}$ /(mol/L)			
$\mid d_i \mid$			
$\bar{d}_r$ /%			

Table 3.24 Determination of aluminum content

	1	2	3
V_2(final)/mL			
V_2(initial)/mL			
V_2/mL			
w_{Al}/%			
$\bar{w}_{Al}$/%			
$\mid d_i \mid$			
$\bar{d}_r$ /%			

Notes

(1) Stir with glass rods to prevent boiling and splashing of solution when it is heating.

(2) The amount of indicator should be moderate.

Questions

(1) Why should the solution be heated to 80～90℃ after adding hexamethylene tetramine buffer solution in the standardization of $CuSO_4$ standard solution?

(2) Is it necessary to accurately add 30 mL of EDTA solution when measuring aluminum content? Why?

(3) What is the purpose of adding NH_4F? What is the impact on the determination if the amount of NH_4F is too much or too little?

3.4 Redox Titration Experiments

Exp.18 Preparation and Standardization of Potassium Permanganate Standard Solution and Determination of H_2O_2

Objectives

(1) To grasp the preparation and standardization of potassium permanganate standard solution.

(2) To master the principle and method for determining hydrogen peroxide by potassium permanganate method.

Principles

The primary standards for the standardization of $KMnO_4$ standard solution are As_2O_3, pure iron wire and $Na_2C_2O_4$, among which $Na_2C_2O_4$ is most commonly used. In an acidic medium, $KMnO_4$ reacts with $Na_2C_2O_4$ as follows:

$$2MnO_4^- + 5C_2O_4^{2-} + 16H^+ = 2Mn^{2+} + 10CO_2\uparrow + 8H_2O$$

No additional indicator is required because the solution itself has a purple-red color.

H_2O_2, also known as hydrogen peroxide, is widely used in the field of industry, biology and medicine. It can be used for bleaching wool, silk fabrics, disinfection and sterilization; pure H_2O_2 can be used as oxidant of rocket fuel; industrially, H_2O_2 can be used to reduce chlorine; biologically, the activity of catalase was measured by utilizing the catalytic effect of catalase on H_2O_2 decomposition reaction.

At room temperature, H_2O_2 can be quantitatively oxidized by $KMnO_4$ solution in dilute H_2SO_4 solution. Therefore, the content of hydrogen peroxide can be determined by $KMnO_4$ method.

$$2MnO_4^- + 5H_2O_2 + 6H^+ = 2Mn^{2+} + 5O_2\uparrow + 8H_2O$$

Materials

Equipment: acid burette (50 mL), Erlenmeyer flask (250 mL), beaker (1000 mL), graduated cylinder (10 mL, 50 mL), volumetric flask (250 mL), pipette (25 mL), analytical balance.

Reagents: $Na_2C_2O_4$ (primary standard), $KMnO_4$ standard solution (0.02 mol/L), H_2SO_4 solution (3 mol/L), $MnSO_4$ solution (1 mol/L), $KMnO_4$ (AR), H_2O_2 sample (commercially available H_2O_2 aqueous solution with a mass fraction of approximately 30%).

Procedures

(1) Preparation of $KMnO_4$ standard solution (0.02 mol/L).

Weigh 1.6 g of $KMnO_4$ and dissolve with 500 mL of distilled water, cover with a watch glass, and boil for 20～30 min by heating (supply a certain amount of distilled water during the heating process to keep the volume of the solution unchanged). After cooling, transfer the solution to a brown bottle and leave it in the dark for 7～10 days (if the solution is boiled and kept on the water bath for 1 h, place it for 2～3 days), then use a microporous glass funnel (No. 3 or No. 4) filter to remove impurities such as MnO_2. The filtrate was stored in a clean brown bottle with stopper and to be standardized.

(2) Standardization of $KMnO_4$ standard solution (0.02 mol/L).

Accurately weigh 0.13～0.16 g of $Na_2C_2O_4$ primary standard in a 250 mL Erlenmeyer flask, add 40 mL of water, 10 mL of H_2SO_4 (3 mol/L). Heat to 70～80℃ (the temperature at which steam starts to be formed), and titrate with $KMnO_4$ solution while it is hot. The titration speed should be slow when approaching the endpoint and shaken well until the solution is reddish and does not fade within 30 s. Note that the temperature of the solution at the endpoint should be kept above 60℃. Take three replicates in parallel. Record the consumed volume of $KMnO_4$ standard solution in Table 3.25.

(3) Determination of hydrogen peroxide.

Pipette 2.00 mL of H_2O_2 sample solution to a 250 mL volumetric flask, dilute to the mark with water and shake well. Transfer 25.00 mL of the diluted solution into a 250 mL Erlenmeyer flask, add 5 mL of H_2SO_4 solution (3 mol/L) and 2～3 drops of $MnSO_4$ solution (1 mol/L), and then titrate with $KMnO_4$ standard solution until the solution is reddish and does not fade within 30 s. Take three replicates in parallel. Record the consumed volume of $KMnO_4$ standard solution in Table 3.26.

Data Recording and Processing

(1) Calculate the concentration of $KMnO_4$ standard solution and relative mean deviation according to the consumed volume of $KMnO_4$ solution for standardization.

(2) Calculate the H_2O_2 content (g/L) in the sample and the relative mean deviation according to the concentration of $KMnO_4$ standard solution, the volume consumed by the titration and the dilution of the sample before titration.

Table 3.25　Standardization of $KMnO_4$ standard solution

	1	2	3
$m_{Na_2C_2O_4}$ /g			
V_{KMnO_4} (final) /mL			
V_{KMnO_4} (initial)/mL			
V_{KMnO_4} /mL			
c_{KMnO_4} /(mol/L)			
$\bar{c}_{KMnO_4}$ /(mol/L)			
$\|d_i\|$			
$\bar{d}_r$ /%			

Table 3.26 Determination of H_2O_2

	1	2	3
$V_{H_2O_2}$ /mL			
V_{KMnO_4} (final) /mL			
V_{KMnO_4} (initial)/mL			
V_{KMnO_4} /mL			
$\rho_{H_2O_2}$ /(g/L)			
$\bar{\rho}_{H_2O_2}$ /(g/L)			
$\lvert d_i \rvert$			
$\bar{d}_r$ /%			

Notes

(1) The reaction rate between potassium permanganate and sodium oxalate is slow at room temperature, therefore the solution must be heated. However, the temperature should not be too high because oxalic acid will decompose if it exceeds 90℃:

$$H_2C_2O_4 = CO_2\uparrow + CO\uparrow + H_2O$$

(2) If the titration rate is too fast, some potassium permanganate will not react with sodium oxalate due to the decomposition in a hot acidic solution:

$$4MnO_4^- + 4H^+ = 4MnO_2 + 3O_2\uparrow + 2H_2O$$

Questions

(1) Under what conditions should the $KMnO_4$ standard solution be standardized with $Na_2C_2O_4$ as a primary standard?

(2) When standardizing the $KMnO_4$ standard solution, why does the red color of the solution fade slowly after the first drop of $KMnO_4$ is added, and then the fade rate is getting faster and faster?

(3) Is it possible to adjust the acidity with HNO_3 or HCl solution when measuring H_2O_2 by potassium permanganate method? Why?

Exp.19 Determination of Iron in Iron Ore by Potassium Dichromate Method (Mercury-Free Method)

Objectives

(1) To grasp the principle and procedure for the determination of iron by mecury-free method.

(2) To understand the green chemistry significance of iron determination by mercury-free method.

Principles

There are many types of iron ores, and magnetite (Fe_3O_4), hematite (Fe_2O_3) and siderite

($FeCO_3$) are mainly used for iron making. After the sample is dissolved in HCl, the iron is converted into Fe^{3+}. First, most of the Fe (III) is reduced to Fe (II) by using $SnCl_2$ in the hot concentrated HCl solution, and the remaining Fe (III) is reduced by $TiCl_3$. The equations are as follows:

$$2Fe^{3+} + SnCl_4^{2-} + 2Cl^- = 2Fe^{2+} + SnCl_6^{2-}$$

$$Fe^{3+} + Ti^{3+} + H_2O = Fe^{2+} + TiO^{2+} + 2H^+$$

After all parts of Fe (III) are quantitatively reduced to Fe (II), a slight excess of $TiCl_3$ can reduce the pretreatment indicator Na_2WO_4 in the solution from colorless to blue W (V) (commonly known as tungsten blue). The excess tungsten blue is then oxidized with a small amount of dilute $K_2Cr_2O_7$ solution so that the blue color just fades out, indicating the endpoint of the pre-reduction.

After pretreatment, the solution is titrated with $K_2Cr_2O_7$ standard solution in the sulfur-phosphorus mixed acid medium with sodium diphenylamine sulfonate as an indicator, until the solution is purple, which is the endpoint.

$SnCl_2$-$TiCl_3$-$K_2Cr_2O_7$ mercury-free method for iron detection avoids environmental pollution by mercury method and has been designated as the national standard for iron ore analysis.

Materials

Equipment: acid burette (50 mL), Erlenmeyer flask (250 mL), graduated cylinder (100 mL), volumetric flask (250 mL), pipette (10 mL, 25 mL), beaker (100 mL), watch glass, analytical balance, etc.

Reagents: $K_2Cr_2O_7$ standard solution (0.017 mol/L), concentrated HCl, $SnCl_2$ solution (50 g/L), $TiCl_3$ solution (15 g/L), sulfur-phosphorus mixed acid solution, Na_2WO_4 solution (250 g/L), aqueous solution of sodium diphenylamine sulfonate (2 g/L), iron ore sample.

Procedures

(1) Preparation of $K_2Cr_2O_7$ standard solution (0.017 mol/L).

Accurately weigh 1.2～1.3 g of $K_2Cr_2O_7$ in a 100 mL beaker, add appropriate amount of water to dissolve, quantitatively transfer it to a 250 mL volumetric flask, dilute to the mark with distilled water, and shake well. Calculate the concentration of $K_2Cr_2O_7$ standard solution.

(2) Sample dissolution.

Weigh about 0.2 g of iron ore sample in a 250 mL Erlenmeyer flask, wet with a small amount of water, add 10 mL of concentrated HCl solution, and add 8～10 drops of $SnCl_2$ solution to help dissolve. Cover with a watch glass and heat it in a nearly boiling water bath for 20～30 min until the residue is white, indicating that the sample is completely dissolved. The color of the final solution is orange-yellow. Rinse the watch glass and the inner wall of the Erlenmeyer flask with a small amount of distilled water.

(3) Sample pre-treatment.

Carefully add $SnCl_2$ solution to the sample solution while it is hot to reduce Fe (III) and shake it while dropping until the solution turns from pale brown to light yellow, indicating that most

Fe (III) has been reduced. Add 4 drops of Na_2WO_4 and 60 mL of water and heat. Add $TiCl_3$ solution drop by drop until a stable light blue solution was observed. Rinse the outer wall of the Erlenmeyer flask with tap water to cool the solution to room temperature. Carefully add 10 times diluted $K_2Cr_2O_7$ solution until blue fades out.

(4) Determination of iron.

The sample solution was diluted to 150 mL with distilled water, mixed with 15 mL of sulfur-phosphorus acid and 5~6 drops of sodium diphenylamine sulfonate as indicator. Then Fe (II) was determined immediately with $K_2Cr_2O_7$ standard solution until the solution was stable purple. Record the consumed volume of $K_2Cr_2O_7$ solution in Table 3.27. Take three replicates in parallel.

Data Recording and Processing

Calculate the mass fraction of iron in iron ore and relative average deviation according to the consumed volume of potassium dichromate standard solution.

Table 3.27 Determination of iron content

	1	2	3
$m_{K_2Cr_2O_7}$ /g			
$c_{K_2Cr_2O_7}$ /(mol/L)			
$V_{K_2Cr_2O_7}$ (final)/mL			
$V_{K_2Cr_2O_7}$ (initial)/mL			
$V_{K_2Cr_2O_7}$ /mL			
c_{Fe} /(mol/L)			
w_{Fe}/%			
$\overline{w}_{Fe}$ /%			
$\|d_i\|$			
$\overline{d}_r$ /%			

Questions

(1) Why should $SnCl_2$ solution be added when the sample solution is hot during the pretreatment procedure?

(2) Why is it necessary to use two reducing agents ($SnCl_2$ and $TiCl_3$) in the prereduction of Fe (III) to Fe (II)? What is the disadvantage of using only one reducing agents?

(3) What is the function of adding H_3PO_4 before titration? Why do we titrate immediately after H_3PO_4 joining?

Exp.20 Determination of Chemical Oxygen Demand in Environmental Water Samples ($KMnO_4$ Method)

Objectives

(1) To preliminarily learn the significance of the environmental analysis and methods of

collecting and preserving water samples.

(2) To understand the definition and representation of chemical oxygen demand (COD) and the significance to determine it.

(3) To master the principle and procedure of the determination of COD in water samples by potassium permanganate method.

Principles

Chemical oxygen demand (COD) is an important index of water quality control. It refers to the amount of the strong oxidants to oxidize the reducing substances in a certain volume of water sample, corresponding to the amount of O_2 (mg/L). Most of the reducing substances in wastewater are organic, thus COD is an important index to evaluate the pollution of water induced by organic substances. The determination of COD can be conducted by potassium permanganate or potassium dichromate method. Among them, the potassium permanganate method is applicable to water sample without serious pollution, such as surface water, drinking water and domestic sewage; the potassium dichromate method applies to industrial wastewater.

In this experiment, the potassium permanganate method is used to determine the COD of water samples. A known amount of potassium permanganate standard solution is first added to a water sample under acidic condition. Complete reaction between potassium permanganate and organic pollutants in sample is achieved by heating. Then the excess potassium permanganate is reacted with a known amount of sodium oxalate standard solution. Finally excess sodium oxalate is titrated using potassium permanganate standard solution. The chemical oxygen demand of the water sample can be calculated based on the consumed amount of potassium permanganate standard solution. The main reactions involved are as follows:

$$4MnO_4^- + 5C + 12H^+ = 4Mn^{2+} + 5CO_2\uparrow + 6H_2O$$

$$2MnO_4^- + 5C_2O_4^{2-} + 16H^+ = 2Mn^{2+} + 10CO_2\uparrow + 8H_2O$$

$$5C_2O_4^{2-} + 2MnO_4^- + 16H^+ = 2Mn^{2+} + 10CO_2\uparrow + 8H_2O$$

Here, C refers to a reducing substance or an aerobic substance in water, mainly an organic substance.

Materials

Equipment: acid burette (50 mL), volumetric flask (250 mL), Erlenmeyer flask (250 mL), beaker (100 mL), analytical balance, etc.

Reagents: $KMnO_4$ standard solution (0.02 mol/L, 0.002 mol/L), $Na_2C_2O_4$ (s), H_2SO_4 solution (6 mol/L), silver sulfate (s).

Procedures

(1) Preparation of $Na_2C_2O_4$ standard solution (0.005 mol/L).

Weigh about 0.17 g of $Na_2C_2O_4$ (dried at 100～105℃ for 2 h) in a 100 mL beaker, dissolve with distilled water, and then transfer it to a 250 mL volumetric flask. Dilute to the mark and mix well.

(2) Preparation and standardization of $KMnO_4$ standard solution (0.002 mol/L).

Accurately transfer 25.00 mL of 0.02 mol/L $KMnO_4$ solution to a 250 mL volumetric flask, dilute to mark with water, shake well and store in dark. Pipette 25.00 mL of $Na_2C_2O_4$ standard solution in a 250 mL Erlenmeyer flask. Add 5 mL of H_2SO_4 solution (6 mol/L) and heat to 75～85℃ on a water bath. Titrate with $KMnO_4$ standard solution (0.002 mol/L) while the solution is still hot (the titration speed goes from slow to fast, then to slow) until the solution is reddish and keeps for 30 s, which is the endpoint. Record the data in Table 3.28. Repeat the titration 3 times.

(3) Determination of COD in water sample.

Accurately transfer 100.00 mL of water sample to a 250 mL Erlenmeyer flask. Add 5 mL of H_2SO_4 solution (6 mol/L) and several pieces of zeolite, add 10.00 mL of $KMnO_4$ standard solution (0.002 mol/L) by the burette. Boil the resulting solution immediately, and then simmer for 10 min from the first large bubble (the red color remains during this period, if the color fades out, supplement a certain amount of 0.002 mol/L $KMnO_4$ standard solution until it is stable red). Record the total added volume of $KMnO_4$ standard solution (V_1) in Table 3.29. Remove the Erlenmeyer flask and cool for 1 min, add 10.00 mL of $Na_2C_2O_4$ standard solution, shake well (the solution turns from red to colorless, otherwise add more $Na_2C_2O_4$ standard solution), and then titrate the resulting solution with $KMnO_4$ standard solution (0.002 mol/L) until it is reddish and keeps for 30 s. Record the consumption of $KMnO_4$ standard solution (V_2) in Table 3.29. Repeat the titration 3 times.

(4) Determination of COD in the blank sample.

To determine the COD in the blank sample, the same amount of distilled water (100.00 mL) is used instead of water sample for the following experiment as the same procedure (3).

Data Recording and Processing

(1) Calculate the concentration of $KMnO_4$ standard solution, the mean value and the relative average deviation ($\bar{d}_r \leqslant 0.2\%$).

(2) Calculate the COD_{Mn} in water sample.

$$\text{COD/(mg/L)} = \frac{\left[\frac{5}{4}c_{KMnO_4} \times (V_1 + V_2) - \frac{1}{2}c_{Na_2C_2O_4} \times V_{Na_2C_2O_4}\right] \times M_{O_2} \times 1000}{V_{\text{water sample}}}$$

$$COD_{Mn} = \overline{COD}_{tot} - \overline{COD}_{blank}$$

Table 3.28 Standardization of $KMnO_4$ standard solution

	1	2	3
V_{KMnO_4} (final)/mL			
V_{KMnO_4} (initial)/mL			
V_{KMnO_4}/mL			
c_{KMnO_4}/(mol/L)			
$\bar{c}_{KMnO_4}$/(mol/L)			
$\lvert d_i \rvert$			
$\bar{d}_r$/%			

Table 3.29 Determination of COD in water sample

	1	2	3
V_1/mL			
V_2/mL			
(V_1+V_2)/mL			
COD_{tot}/(mg/L)			
$\overline{COD}_{tot}$/(mg/L)			
$\|d_i\|$			
$\bar{d}_r$/%			
COD_{blank}/(mg/L)			
$\overline{COD}_{blank}$/(mg/L)			
COD_{Mn}/(mg/L)			

Notes

(1) The presence of chloride ions can cause interference when its concentration is over 300 mg/L. It can be reduced or eliminated by lowering the concentration of chloride ions through dilution. Otherwise, it can be prevented by the addition of silver sulphate.

(2) Immediate analysis should be made after sampling. Otherwise, add a small amount of sulfuric acid to make the solution pH<2, thus inhibit the decomposition of organic matter by microorganisms, store at 0～5℃, and measure within 48 h.

(3) The solution should remain red when it is boiled. Otherwise, it means that there are more organic substances in the water, and an appropriate amount of potassium permanganate standard solution should be added.

(4) The amount of water sample can be judged initially: 100 mL of clean and transparent water sample; if the water sample is seriously polluted and turbid, take 10～30 mL and diluted to 100 mL with distilled water.

Questions

(1) What is the interference if the chloride ion content is high in the water sample? How to reduce or eliminate its interference?

(2) What is the color of the sample solution after it mixed with potassium permanganate and boiled? What happened if it is colorless and what should be done?

(3) Why is back titration used for this experiment?

(4) What factors can affect the result of COD? Why?

Exp.21 Determination of Chemical Oxygen Demand in Sewage ($K_2Cr_2O_7$ Method)

Objectives

(1) To learn the significance and method for the determination of COD.

(2) To master the principle and procedure for determination of COD by potassium dichromate method.

Principles

Chemical oxygen demand (COD) is an indicator of the reductive substances in water. It can reflect the pollution of water bodies by organic matter. The principle for the determination of COD by the potassium dichromate method in water samples is as follow: the water sample is firstly mixed with a known excess amount of potassium dichromate standard solution under strong acidic condition. The mixture is heated and refluxed, where the reductive matter in water sample is oxidized by potassium dichromate. The remaining potassium dichromate is titrated with ammonium ferrous sulfate standard solution using ferrion as indicator. The value of COD can be calculated according to the amount of potassium dichromate consumed in the titration. The reaction involved is as follows:

$$6Fe(NH_4)_2(SO_4)_2 + K_2Cr_2O_7 + 7H_2SO_4 = 3Fe_2(SO_4)_3 + Cr_2(SO_4)_3 + K_2SO_4 + 6(NH_4)_2SO_4 + 7H_2O$$

Silver sulfate can be added as a catalyst in the experiment, and acidic potassium dichromate can oxidize most of the organic compounds including linear aliphatic compounds. Chloride ions can be oxidized by dichromate and react with silver sulfate to cause precipitation, which affects the measurement results. Therefore, mercury sulfate can be added to the water sample before reflux to complex the chloride ions to eliminate interference. Samples with a chloride ion content higher than 1000 mg/L should be diluted first and then measured.

Materials

Equipment: acid burette (50 mL), pipette (10 mL), reflux device with an Erlenmeyer flask (250 mL), beaker (200 mL), analytical balance, electric furnace, etc.

Reagents: ferrion indicator solution [1.485 g of *o*-phenanthroline ($C_{12}H_8N_2 \cdot H_2O$) and 0.695 g of $FeSO_4 \cdot 7H_2O$ are dissolved and diluted to 100 mL with water, the solution is stored in brown reagent bottle], potassium dichromate standard solution (0.25 mol/L), ammonium ferrous sulfate standard solution (0.1 mol/L), mercury sulfate (s), silver sulfate (s), sulfuric acid solution (6 mol/L), water sample.

Procedures

(1) Preparation of potassium dichromate standard solution (0.25 mol/L).

Weigh about 18.4 g of $K_2Cr_2O_7$ (dried at 120℃ for 2 h), dissolve it in water, and then transfer it to a 250 mL volumetric flask. Dilute to mark with water and shake well.

(2) Preparation and standardization of ammonium ferrous sulfate standard solution (0.1 mol/L).

Weigh about 9.8 g of ammonium ferrous sulfate in a 250 mL beaker, dissolve it with water, and then slowly add 20 mL of sulfuric acid (6 mol/L) under stirring. After dissolving, dilute to 250 mL with water and store in a reagent bottle.

Accurately pipette 10.00 mL of potassium dichromate standard solution to a 250 mL Erlenmeyer flask. Dilute it with 30 mL of distilled water, and slowly add 30 mL of sulfuric acid (6 mol/L), shake well. After cooling, add 3 drops of ferrion indicator. Titrate the solution with ammonium ferrous sulfate standard solution until the color of the solution turns from yellow to bluish green to reddish brown. Record the consumed volum of $Fe(NH_4)_2(SO_4)_2$ in Table 3.30. Repeat the titration 3 times.

(3) Determination of COD in water samples.

Accurately transfer 20.00 mL of the homogeneous water sample (or dilute the appropriate amount of water to 20.00 mL) to a 250 mL Erlenmeyer flask. Add 0.4 g of mercury sulfate, 10.00 mL of potassium dichromate standard solution (0.25 mol/L) and several pieces of zeolite. A reflux condenser is attached on the flask. Add about 30 mL of sulfuric acid-silver sulfate solution from the opening of the condenser. Mix the solution homogeneously by smoothly shaking, and then refluxed it for 2 h (starting at the time of boiling). After cooling, the condenser is washed with 90 mL of water. The total volume of the solution should be less than 140 mL. Remove the Erlenmeyer flask. After the solution is cooled again, add three drops of ferrion indicator. The solution is titrated with ammonium ferrous sulfate standard solution until the color of the solution changes from yellow to bluish green to reddish brown. Record the consumed volume of ammonium ferrous sulfate standard solution in Table 3.31. Repeat the titration 3 times.

(4) Determination of COD in the blank sample.

To determine the blank, the same amount of distilled water (20.00 mL) is used instead of water sample for the following experiment in the same procedure.

Data Recording and Processing

(1) Calculate the concentration of potassium dichromate standard solution.

(2) Calculate the concentration of ammonium ferrous sulfate standard solution, the mean value and the relative average deviation ($\bar{d}_r \leqslant 0.2\%$).

$$c = \frac{10.00 \times 6 \times c_{K_2Cr_2O_7}}{V}$$

Where, c is the concentration of the ammonium ferrous sulfate standard solution, mol/L; V is the volume of the consumed ammonium ferrous sulfate standard solution, mL.

(3) Calculate the COD_{Cr} in water sample.

$$COD_{Cr}\text{(represented by } O_2\text{, mg/L)} = \frac{(V_1 - V_0) \times c \times 8 \times 1000}{V}$$

Where, c is the concentration of ammonium ferrous sulfate standard solution, mol/L; V_0 is the volume of ammonium ferrous sulfate consumed for the sample, mL; V_1 is the volume of ammonium ferrous sulfate consumed for the blank, mL; V is the volume of water sample, mL; 8 is the half molar mass of O, g/mol.

Table 3.30　Standardization of ammonium ferrous sulfate standard solution

	1	2	3
$V_{Fe(NH_4)_2(SO_4)_2}$ (final)/mL			
$V_{Fe(NH_4)_2(SO_4)_2}$ (initial)/mL			

Continued

	1	2	3
$V_{Fe(NH_4)_2(SO_4)_2}$ /mL			
$c_{Fe(NH_4)_2(SO_4)_2}$ /(mol/L)			
$\bar{c}_{Fe(NH_4)_2(SO_4)_2}$ /(mol/L)			
$\lvert d_i \rvert$			
$\bar{d}_r$/%			

Table 3.31 Determination of COD in water sample

	1	2	3
V_0/mL			
V_1/mL			
(V_1-V_0)/mL			
COD_{Cr} /(mg/L)			
$\overline{COD_{Cr}}$/(mg/L)			
$\lvert d_i \rvert$			
$\bar{d}_r$/%			

Notes

(1) The volume of water sample required is in the range of 10.00～50.00 mL, and the amount and concentration of reagents should be changed according to Table 3.32.

Table 3.32 Table of required amount of reagents depending on the volume of sample

Sample volume/mL	0.25 mol/L potassium dichromate/mL	Sulfuric acid-silver sulfate/mL	Ammonium ferrous sulfate/(mol/L)	Mercury sulfate/g	Total volume before titration/mL
10.0	5.0	15	0.050	0.2	70
20.0	10.0	30	0.100	0.4	140
30.0	15.0	45	0.150	0.6	210
40.0	20.0	60	0.200	0.8	280
50.0	25.0	75	0.250	1.0	350

(2) For the water sample with the COD value lower than 50 mg/L, 0.0250 mol/L potassium dichromate standard solution should be used and 0.01 mol/L ammonium ferrous sulfate standard solution should be used for back titration.

(3) The amount of remaining potassium dichromate in solution after refluxing should be 1/5～4/5 of the original amount.

(4) The concentration of ammonium ferrous sulfate solution should be standardized in each experiment, pay particular attention to changes in concentration when room temperature is high.

Questions

(1) Why is blank experiment required?

(2) What factors will affect the determination of COD_{Cr}? How to eliminate the interference?

(3) What should be noticed while using ammonium ferrous sulfate standard solution in experiment?

Exp.22 Determination of Copper in Copper Salt by Indirect Iodimetry

Objectives

(1) To master the principle of the determination of copper by indirect iodometry.

(2) To learn about the source of error in indirect iodometry.

Principles

Under weakly acidic conditions (pH 3～4), Cu^{2+} can be reduced to CuI by KI, and at the same time, I_2 is precipitated. With starch as indicator, the content of copper can be analyzed by titrating the precipitated iodine with $Na_2S_2O_3$ standard solution. The reactions are as follows:

$$2Cu^{2+} + 4I^- = 2CuI\downarrow + I_2$$

or

$$2Cu^{2+} + 5I^- = 2CuI\downarrow + I_3^-$$

$$2S_2O_3^{2-} + I_2 = S_4O_6^{2-} + 2I^-$$

Here, I^- is not only employed as a reducing agent of Cu^{2+}, but also as a precipitator of Cu^+ and a complexing agent of I_2.

Indirect iodometry must be carried out in a weakly acidic or neutral solution. For the determination of Cu^{2+}, the acidity of the solution is usually controlled to pH 3～4 using a NH_4HF_2 buffer solution (i.e., HF/F^- conjugated acid-base pair). In addition, the F^- in NH_4HF_2 buffer solution can be employed as the masking agent, converting the coexistent Fe^{3+} into FeF_6^{3-} to eliminate its interference. If the sample does not contain Fe^{3+}, the acidity can also be controlled with an acetic acid buffer solution (pH ≈ 4).

The surface of CuI precipitation is easy to adsorb a small amount of I_2, and the adsorbed I_2 is difficult to interact with starch, so the endpoint is advanced. Generally, the KSCN solution should be added near the endpoint to convert the CuI precipitate into CuSCN with smaller solubility. Thus the I_2 adsorbed by CuI can be released, improving the accuracy of the determination because CuSCN does not adsorb I_2.

Materials

Equipment: acid burette (50 mL), volumetric flask (250 mL), Erlenmeyer flask (250 mL), iodine volumetric flask (250 mL), analytical balance.

Reagents: $Na_2S_2O_3$ standard solution (0.1 mol/L), $K_2Cr_2O_7$ standard solution (0.017 mol/L),

KI solution (100 g/L, freshly prepared before use), KSCN solution (100 g/L), H_2SO_4 solution (1 mol/L), HCl solution (6 mol/L), starch solution (5 g/L), sample of $CuSO_4 \cdot 5H_2O$.

Procedures

(1) Preparation and Standardization of $Na_2S_2O_3$ standard solution (0.1 mol/L).

Weigh 13 g of $Na_2S_2O_3$ and dissolve in 250 mL of freshly boiled and cooled distilled water. After dissolving, add about 0.1 g of Na_2CO_3, dilute to 500 mL with freshly boiled and cooled distilled water, store in brown reagent bottle, and place in dark place. Standardize the solution after 3～5 days.

Accurately transfer 25.00 mL of $K_2Cr_2O_7$ standard solution (0.017 mol/L) to an Erlenmeyer flask, add 5 mL of HCl solution (6 mol/L), 5 mL of KI solution (100 g/L), shake well, and place in the dark for 5 min to complete the reaction. Add 50 mL of distilled water, titrate to pale yellow with the $Na_2S_2O_3$ standard solution to be standardized, then add 3 mL of starch solution (5 g/L) and continue titrating until the solution is bright green. Record the consumed volume of $Na_2S_2O_3$ standard solution in Table 3.33. Repeat the titration 3 times.

(2) Determination of copper content in copper salt.

Accurately weigh 0.5～0.6 g of $CuSO_4 \cdot 5H_2O$ in a 250 mL Erlenmeyer flask. Add 5 mL of 1 mol/L H_2SO_4 solution and 100 mL of water to dissolve. Add 10 mL of KI solution (100 g/L) and titrate with the $Na_2S_2O_3$ standard solution immediately until the color of the solution is light yellow. Add 2 mL of 5 g/L starch solution and continue to titrate until the color of the solution is light blue. Then add 10 mL of KSCN solution (100 g/L), the color of the solution is dark blue at this time. Titrate with $Na_2S_2O_3$ standard solution until the blue color fades out, which is the endpoint. Record the consumed volume of $Na_2S_2O_3$ standard solution in Table 3.34. Repeat the titration 3 times.

Data Recording and Processing

Calculate the concentration of $Na_2S_2O_3$, calculate the mass fraction of copper in sample, the mean value and the relative average deviation ($\bar{d}_r \leqslant 0.2\%$).

Table 3.33 Standardization of $Na_2S_2O_3$ standard solution

	1	2	3
$V_{Na_2S_2O_3}$ (final)/mL			
$V_{Na_2S_2O_3}$ (initial)/mL			
$V_{Na_2S_2O_3}$ /mL			
$c_{Na_2S_2O_3}$/ (mol/L)			
$\bar{c}_{Na_2S_2O_3}$ /(mol/L)			
$\lvert d_i \rvert$			
$\bar{d}_r$ /%			

Table 3.34 Determination of Cu content

	1	2	3
$V_{Na_2S_2O_3}$ (final)/mL			
$V_{Na_2S_2O_3}$ (initial)/mL			
$V_{Na_2S_2O_3}$ /mL			
c_{Cu}/(mol/L)			
w_{Cu}/%			
$\overline{w}_{Cu}$/%			
$\mid d_i \mid$			
$\overline{d}_r$ /%			

Notes

(1) Keep shaking the sample solution during the titration.

(2) The pH of the solution should be generally controlled at 3～4. If the acidity is too low, it is easy for Cu^{2+} to hydrolyze; if the acidity is too high, I^- will be oxidized by O_2 in the air.

Questions

(1) What is the function of KI in the experiment?

(2) Why should KSCN be added? Why cannot add it early?

(3) What reagent can be used to eliminate the interference of iron and control the pH to 3～4 simultaneously in the titration of sample with iron?

Exp.23 Determination of Calcium in Calcium Preparations

Objectives

(1) To master the principle and method of determining calcium by $KMnO_4$ method.

(2) To understand the basic requirements and operation of precipitation separation.

Principles

Some metal ions (such as Pb^{2+}, Cd^{2+}, etc.) can combine with $C_2O_4^{2-}$ to form insoluble oxalate precipitates. The precipitate is filtered, washed, and dissolved with H_2SO_4 solution, and then the released $H_2C_2O_4$ could be titrated with $KMnO_4$ standard solution. Therefore, the metal ions can be determined indirectly. Taking Ca^{2+} as an example, the reactions are as follows:

$$Ca^{2+} + C_2O_4^{2-} = CaC_2O_4 \downarrow$$

$$CaC_2O_4 + 2H^+ = H_2C_2O_4 + Ca^{2+}$$

$$5H_2C_2O_4 + 2MnO_4^- + 6H^+ = 2Mn^{2+} + 10CO_2\uparrow + 8H_2O$$

This method can be used to determine the calcium levels in some calcium preparations (such as calcium gluconate, calcium tablet, etc.).

Materials

Equipment: acid burette (50 mL), pipette (25 mL), volumetric flask (250 mL), Erlenmeyer flask (250 mL), beaker (100 mL), analytical balance, hot plate.

Reagents: $KMnO_4$ standard solution (0.02 mol/L), $(NH_4)_2C_2O_4$ solution (0.05 mol/L), $NH_3 \cdot H_2O$(7 mol/L), HCl solution (6 mol/L), H_2SO_4 solution (1 mol/L), methyl orange aqueous solution (1 g/L), $AgNO_3$ solution (0.1 mol/L), calcium preparations.

Procedures

Weigh 0.05 g of calcium preparations in a 100 mL beaker. Add appropriate amount of distilled water and 2～5 mL of HCl solution (6 mol/L), gently shake the beaker, and heat to dissolve. After cooling, join in the solution 2～3 drops of methyl orange (1 g/L), add $NH_3 \cdot H_2O$ solution (7 mol/L) drop by drop to the solution until the color of the solution changes from red to yellow. Then drop about 50 mL of $(NH_4)_2C_2O_4$ solution (0.05 mol/L), the solution is aged for 30 min on low temperature hot plate (or water bath). Filter the solution after cooling, wash the precipitate in the beaker for several times, and then the precipitate was transferred into the funnel and continued to wash until no Cl^- is detected (the cleaning solution was tested by $AgNO_3$ in HNO_3). Place the filter paper with precipitation on the inner wall of the original beaker, wash the precipitate from the filter paper into the beaker using 50 mL of H_2SO_4 solution (1 mol/L). Then wash with distilled water twice. Finally add distilled water to dilute to 100 mL, and heat to 70～80℃. The solution is titrated with $KMnO_4$ standard solution (0.02 mol/L, see details in Exp. 18) to a reddish color. Then the filter paper is stirred into the solution. If the color of the solution fades out, continue to titrate until the occurrence of pale red and it does not fade within 30 s. Record the consumed volume of $KMnO_4$ standard solution in Table 3.35. Repeat the sample analysis 3 times.

Data Recording and Processing

Calculate the mass fraction of Ca in calcium preparation based on the recorded data.

Table 3.35 Determination of Ca content in calcium preparations

	1	2	3
m_s/g			
V_{KMnO_4} (final)/mL			
V_{KMnO_4} (initial)/mL			
V_{KMnO_4}/mL			
w_{Ca}/%			
$\bar{w}_{Ca}$/%			
$\|d_i\|$			
$\bar{d}_r$/%			

Questions

(1) Why should the $(NH_4)_2C_2O_4$ solution be added drop by drop in the hot solution?

(2) Why should the CaC_2O_4 precipitate be washed till no Cl^- detected?

(3) Try to compare the $KMnO_4$ method and complexometric titration method for the determination of Ca^{2+}.

Exp.24 Determination of Dissolved Oxygen in Water

Objectives

(1) To understand the definition of dissolved oxygen and the significance to determine it.

(2) To master the principle and method for determination of dissolved oxygen in water sample by iodometry.

Principles

Pollution of water with aerobic materials decreases the level of dissolved oxygen (DO), leading to the death of fish under the anaerobic condition. Thus, the DO concentration can reflect the degree of water pollution, and it is an important index of water quality. In this experiment, iodimetry is used to determine the DO concentration. Manganese sulfate and alkaline potassium iodide are added into water sample, and the precipitate of manganese hydroxide is produced. $Mn(OH)_2$ is unstable, which is quickly oxidized by DO to generate a higher valence manganese compound. The reactions are as follows:

$$MnSO_4 + 2NaOH = Mn(OH)_2\downarrow + Na_2SO_4$$

$$2Mn(OH)_2 + O_2 = 2H_2MnO_3$$

$$H_2MnO_3 + Mn(OH)_2 = Mn_2O_3\downarrow + 2H_2O$$

The higher valence manganese compound will oxidize potassium iodide to generate a quantity of free iodine equivalent to DO under acidic condition. The free iodine is titrated with sodium thiosulfate standard solution, and the DO concentration can be calculated. The reactions are as follows:

$$2KI + H_2SO_4 = 2HI + K_2SO_4$$

$$Mn_2O_3 + 2H_2SO_4 + 2HI = 2MnSO_4 + I_2 + 3H_2O$$

$$I_2 + 2Na_2S_2O_3 = 2NaI + Na_2S_4O_6$$

Materials

Equipment: basic burette (50 mL), pipette (25 mL), volumetric flask (1000 mL), iodometric flask (250 mL), Erlenmeyer flask (250 mL).

Reagents: H_2SO_4 solution(1 ∶ 5, 1 volume of concentrated sulfuric acid is added slowly to a beaker with 5 volumes of distilled water with stirring), manganese sulfate solution(weigh 480 g of $MnSO_4{\cdot}4H_2O$ or 364 g of $MnSO_4{\cdot}H_2O$ in distilled water, filter and dilute to 1000 mL), alkaline potassium iodide solution [dissolve 500 g of NaOH in 300～400 mL of distilled water and 150 g of KI (or 135 g of NaI) in 200 mL of distilled water, respectively. After the NaOH solution is cooling, mix the two solutions together, dilute to 1000 mL with distilled water. Stand for 24 h, collect the upper clear liquid, store in a brown bottle with a rubber stopper. Keep it away from light], starch solution (1%, weigh 1 g of soluble starch, add a little water to form a paste, and then dilute to 100 mL with freshly boiled water), potassium dichromate standard solution (0.025 mol/L), sodium thiosulfate solution.

Procedures

(1) Preparation of potassium dichromate standard solution (0.025 mol/L).

Weigh 1.84 g of $K_2Cr_2O_7$ (dried at 105～110℃ for 2 h), dissolve with distilled water, and then transfer to a 250 mL volumetric flask. Dilute to mark with water and shake well.

(2) Preparation and standardization of sodium thiosulfate solution (0.025 mol/L).

Weigh 6.2 g of sodium thiosulfate ($Na_2S_2O_3 \cdot 5H_2O$), dissolve with 1000 mL of distilled water, and then add 0.4 g of NaOH or 0.2 g of Na_2CO_3. The solution is stored in a brown bottle and its concentration is about 0.025 mol/L.

Accurate concentration can be standardized as follows: add 100 mL of distilled water and 1 g of KI in a 250 mL iodometric flask, pipette 10.00 mL of $K_2Cr_2O_7$ standard solution (0.025 mol/L), add 5 mL of H_2SO_4 solution (1 : 5), then stopper the flask and shake well. Keep it in the dark for 5 min, and then titrate with sodium thiosulfate standard solution until the color of the solution changes from brown to light yellow. Add 1 mL of the starch solution, continue titrating until the blue color just fades. Record the consumed volume of $Na_2S_2O_3$ standard solution in Table 3.36.

(3) Collection of the water sample.

Rinse the DO bottle with water first. Then, the sample is injected directly along the inwall of the bottle or along a narrow tube pushed into the bottom of bottle by Syphon method until the volume of the overflow water is about 1/3～1/2 of the bottle. Do not aerate the sample or remain air bubbles in the DO bottle.

(4) Fix of dissolved oxygen.

Pipette 1.00 mL of $MnSO_4$ solution to the DO bottle filled with water sample. Insert the pipette tip under the liquid surface while transferring. Then, add 2 mL of alkaline KI solution. Stopper the bottle, shake to mix the sample, and then stand. When the precipitation drops to half of the bottle, turn over the bottle and shake again until the precipitate drops to the bottom of the bottle. Generally, fix of the dissolved oxygen is conducted in the field sampling.

(5) Precipitation of iodine.

Gently open the stopper. Immediately add 2.00 mL of concentrated H_2SO_4 with the pipette tip inserted under the liquid surface. Close the stopper, mix thoroughly until the precipitate is completely dissolved. Put it in dark for 5 min.

(6) Determination of the sample.

Transfer 100.00 mL of the solution in a 250 mL Erlenmeyer flask. Titrate with $Na_2S_2O_3$ standard solution (0.025 mol/L), until the solution is pale yellow. Add 1 mL of starch solution and continue to titrate until the blue color just fades out. Record the consumed amount of sodium thiosulfate standard solution in Table 3.37. Repeat the titration 3 times.

Data Recording and Processing

(1) Calculate the concentration of sodium thiosulfate standard solution, the mean value and the relative average deviation ($\bar{d}_r \leqslant 0.2\%$).

$$c = \frac{10.00 \times 6 \times m}{V \times M \times 0.25}$$

Where, V is the consumed volume of sodium thiosulfate standard solution, mL; m is the weight of $K_2Cr_2O_7$, g; M is the molar mass of $K_2Cr_2O_7$, g/mol.

(2) Calculate the content of DO.

$$c_{O_2} = \frac{c \times V \times 8 \times 1000}{100}$$

Where, c_{O_2} is the DO concentration in water sample, mg/L; c is the concentration of sodium thiosulfate standard solution, mol/L; V is the consumed volume of sodium thiosulfate standard solution, mL; 8 is the half molar mass of O, g/mol; 100 is the volume of water sample, mL.

Table 3.36 Standardization of $Na_2S_2O_3$ standard solution

	1	2	3
$V_{Na_2S_2O_3}$ (final)/mL			
$V_{Na_2S_2O_3}$ (initial)/mL			
$V_{Na_2S_2O_3}$ /mL			
$c_{Na_2S_2O_3}$/ (mol/L)			
$\bar{c}_{Na_2S_2O_3}$ /(mol/L)			
$\lvert d_i \rvert$			
$\bar{d}_r$ /%			

Table 3.37 Determination of dissolved oxygen

	1	2	3
$V_{Na_2S_2O_3}$ (final)/mL			
$V_{Na_2S_2O_3}$ (initial)/mL			
$V_{Na_2S_2O_3}$ /mL			
c_{O_2} /(mg/L)			
$\bar{c}_{O_2}$ /(mg/L)			

Notes

(1) If water sample contains nitrite interference, sodium azide can be added in alkaline potassium iodide solution to decompose of nitrite to eliminate its interference.

(2) If water sample contains oxidizing substances (such as free chlorine, etc.), a considerable amount of sodium

thiosulfate should be added in advance to eliminate the interference.

(3) If the concentration of ferric iron in water sample is 100~200 mg/L, 1 mL of 40% potassium fluoride solution can be added to eliminate the interference.

Questions

(1) Why is the starch solution added after the solution is titrated to light yellow? Can it be added while the solution is still the dark yellow?

(2) What factors will affect the results in determination of DO by iodimetry? What interference will occur, if the sample contains oxidizing substances, algae or suspended solids?

Exp.25 Determination of Vitamin C in Juice

Objectives

(1) To learn the preparation and standardization of iodine standard solution.

(2) To master the principle and method for determination of Vitamin C by iodimetry.

Principles

Vitamin C (Vc, also called ascorbic acid, $C_6H_8O_6$) can quantitatively react with I_2 because its reduction group of enediol can be oxidized into the group of diketone by I_2. Vitamin C can be titrated by I_2 solution using starch solution as the indicator. The reaction is as follows:

$$C_6H_8O_6 + I_2 = C_6H_6O_6 + 2HI$$

Because of its strong reducibility, Vc is easily oxidized in the air, especially in the alkaline medium. Thus, acetic acid solution can be added during the titration to keep a weak acidic condition, in which the oxidation speed of Vc is much slower.

Materials

Equipment: basic burette (50 mL), pipette (25 mL), Erlenmeyer flask (250 mL), mortar, brown reagent bottle (250 mL).

Reagents: $Na_2S_2O_3 \cdot 5H_2O(s)$, I_2 standard solution (0.05 mol/L), starch solution (5%, weigh 0.5 g of soluble starch, make a paste with a small amount of water, and then slowly add to 100 mL of boiling distilled water, and continue to boil until the solution is transparent), HAc solution (2 mol/L), juice.

Procedures

(1) Preparation and standardization of $Na_2S_2O_3$ standard solution (0.05 mol/L).

See details in Exp. 22.

(2) Preparation and standardization of I_2 standard solution (0.05 mol/L).

Weigh 3.3 g of I_2 and 5 g of KI in a mortar, grind them with small amount of water. After dissolving totally, transfer all the solution to a brown bottle, add water to 250 mL, mix well and store it in the dark. Accurately transfer 25.00 mL of $Na_2S_2O_3$ standard solution to a 250 mL Erlenmeyer flask, add 50 mL of distilled water and 2 mL of starch solution. Then titrate with I_2 standard solution until the color of the solution is blue and keeps for 30 s. Record the consumed volume of I_2 solution in Table 3.38. Repeat the titration 3 times.

(3) Determination of vitamin C in juice beverage.

Accurately transfer 25.00 mL of juice to a 250 mL Erlenmeyer flask. Add 10 mL of HAc solution (2 mol/L) and 2 mL of starch solution, mix well. Then titrate with I_2 solution until the color of the solution is stable blue. Record the consumed volume of I_2 solution in Table 3.39. Repeat the titration 3 times.

Data Recording and Processing

(1) Calculate the concentration of I_2 standard solution, the mean value and the relative average deviation.

(2) Calculate the concentration of Vc in juice.

$$\rho_{Vc} = \frac{cV_1}{V} \times 176.12 \times 1000$$

Where, ρ_{Vc} is the concentration of Vc in juice, mg/L; c is the concentration of I_2 standard solution, mol/L; V_1 is the consumed volume of I_2 standard solution, mL; V is the volume of juice, mL; 176.12 is the molar mass of Vc, g/mol.

Table 3.38　Standardization of I_2 standard solution

	1	2	3
V_{I_2} (final)/mL			
V_{I_2} (initial)/mL			
V_{I_2} /mL			
c_{I_2} /(mol/L)			
$\bar{c}_{I_2}$ /(mol/L)			
$\lvert d_i \rvert$			
$\bar{d}_r$ /%			

Table 3.39　Determination of vitamin C

	1	2	3
V_{I_2} (final)/mL			
V_{I_2} (initial)/mL			
V_{I_2} /mL			
ρ_{Vc} /(g/L)			
$\bar{\rho}_{Vc}$ /(g/L)			
$\lvert d_i \rvert$			
$\bar{d}_r$ /%			

Notes

(1) The titration speed should be fast because vitamin C is easy to be oxidized.

(2) If the juice is viscous, filtration and dilution are necessary before determination.

Questions

(1) Why is HAc solution added in the sample?

(2) What is the function of KI in the preparation of I_2 solution?

Exp.26 Determination of Glucose in Glucose Injection

Objectives

To understand the principle and method of determining glucose by iodometric method.

Principles

I_2 can be disproportionated into IO^- and I^- in an alkaline solution, and IO^- can quantitatively oxidize glucose ($C_6H_{12}O_6$) to gluconic acid ($C_6H_{12}O_7$), and IO^- which does not interact with $C_6H_{12}O_6$ is further disproportionated to IO_3^- and I^-. After acidification of the solution, IO_3^- and I^- can form I_2 precipitate, which is titrated with $Na_2S_2O_3$ standard solution, thereby the content of $C_6H_{12}O_6$ can be calculated. The involved reactions are as follows:

$$I_2 + 2OH^- = IO^- + I^- + H_2O$$

$$C_6H_{12}O_6 + IO^- = I^- + C_6H_{12}O_7$$

The total reaction formula is:

$$I_2 + C_6H_{12}O_6 + 2OH^- = C_6H_{12}O_7 + 2I^- + H_2O$$

The disproportionation of unreacted IO^- occurs under alkaline conditions:

$$3IO^- = IO_3^- + 2I^-$$

Under acidic conditions:

$$IO_3^- + 5I^- + 6H^+ = 3I_2 + 3H_2O$$

which is

$$IO^- + I^- + 2H^+ = I_2 + H_2O$$

$$I_2 + 2S_2O_3^{2-} = 2I^- + S_4O_6^{2-}$$

It can be seen from the above reaction that one molecule of glucose is equivalent to one molecule of I_2.

Materials

Equipment: acid burette (50 mL), volumetric flask (250 mL), Erlenmeyer flask (250 mL), analytical balance.

Reagents: HCl solution (2 mol/L), NaOH solution (0.2 mol/L), $Na_2S_2O_3$ standard solution (0.05 mol/L), I_2 standard solution (0.05 mol/L), starch solution (0.5%), KI (s), glucose injection (0.5%, 10-fold dilution of 5% glucose injection).

Procedures

(1) Preparation and standardization of $Na_2S_2O_3$ standard solution (0.05 mol/L).

See details in Exp. 22.

(2) Preparation and standardization of I_2 standard solution (0.05 mol/L).

See details in Exp. 25.

(3) Determination of glucose content.

Pipette 25.00 mL of glucose injection to a 250 mL volumetric flask, dilute to the mark and shake well. Pipette 25.00 mL of diluted glucose solution to a 250 mL Erlenmeyer flask, accurately add 25.00 mL of I_2 standard solution, slowly add 0.2 mol/L NaOH, and shake while adding until the solution is light yellow (the speed of adding alkali should not be too fast, otherwise the generated IO^- can not oxidize $C_6H_{12}O_6$, resulting in a lower result). Cover the Erlenmeyer flask with a watch glass, stand for 10~15 min, then add 6 mL of HCl solution (2 mol/L) to make the solution be acidic, and titrate with $Na_2S_2O_3$ standard solution immediately. When the solution is light yellow, add 3 mL of starch solution and continue to titrate until the blue color just fades out. Three portions were titrated in parallel. Record the consumed volume of $Na_2S_2O_3$ solution in Table 3.40.

Data Recording and Processing

Calculate the glucose content in the sample according to the recorded experimental data.

Table 3.40 Determination of glucose content in glucose injection

	1	2	3
$V_{Na_2S_2O_3}$ (final)/mL			
$V_{Na_2S_2O_3}$ (initial)/mL			
$V_{Na_2S_2O_3}$ /mL			
ρ /(g/mL)			
$\bar{\rho}$ /(g/mL)			
$\lvert d_i \rvert$			
$\bar{d}_r$ /%			

Questions

(1) Why is KI added when preparing I_2 solution?

(2) What are the main sources of error in the iodometric method? How to eliminate the errors?

3.5 Precipitation Titration and Gravimetric Analysis Experiments

Exp.27 Determination of Chlorine in Soluble Chloride by Mohr Method

Objectives

(1) To master the preparation and standardization of $AgNO_3$ standard solution.

(2) To master the principle and procedures of Mohr method.

Principles

In neutral or alkaline solution, Cl^- in the test solution is titrated by $AgNO_3$ standard solution using K_2CrO_4 as indicator. The solubility of AgCl is smaller than that of Ag_2CrO_4. Therefore, when all the chloride ions are precipitated, excessive Ag^+ reacts with CrO_4^{2-} and a brick red precipitate of Ag_2CrO_4 generates, indicating the endpoint. The main reactions are as follows:

$$Ag^+ + Cl^- = AgCl\downarrow \text{ (white)} \qquad K_{sp} = 1.8 \times 10^{-10}$$

$$2Ag^+ + CrO_4^{2-} = Ag_2CrO_4\downarrow \text{ (brick red)} \qquad K_{sp} = 2.0 \times 10^{-12}$$

The amount of chlorine in the sample can be calculated based on the consumed volume and concentration of $AgNO_3$ standard solution.

Materials

Equipment: acid burette (50 mL), pipette (25 mL), volumetric flask (100 mL, 250 mL), Erlenmeyer flask (250 mL), analytical balance, brown reagent bottle (500 mL).

Reagents: $AgNO_3$ (AR), NaCl (GR, dried at 500～600℃ for 2～3 h and stored in a desiccator), K_2CrO_4 solution (50 g/L), NaCl sample.

Procedures

(1) Preparation and standardization of $AgNO_3$ standard solution (0.1 mol/L).

Weigh 8.5 g of $AgNO_3$ in a beaker, dissolve with distilled water. Transfer the mixed solution to a brown reagent bottle, dilute to 500 mL. Stopper the bottle and store in the dark.

Accurately weigh 0.55 g～0.60 g of NaCl primary standard in a small beaker, dissolve with distilled water and transfer to a 100 mL volumetric flask. Then dilute to mark with water and mix well. Accurately transfer 25.00 mL of NaCl solution to a 250 mL Erlenmeyer flask, add 20 mL of distilled water and 1 mL of K_2CrO_4 solution (50 g/L). Titrate with $AgNO_3$ standard solution until the color of the solution is brick red. Record the consumed volume of $AgNO_3$ standard solution in Table 3.41. Repeat the titration 3 times.

(2) Determination of chlorine in the sample.

Weigh 1.6 g of NaCl sample in a small beaker, dissolve with distilled water, transfer to a 250 mL volumetric flask, dilute to the mark and mix well. Transfer 25.00 mL of the solution to a

250 mL Erlenmeyer flask. Add 20 mL of distilled water and 1 mL of K_2CrO_4 solution (50 g/L). Titrate with $AgNO_3$ standard solution until the color of the solution is brick red. Record the consumed volume of $AgNO_3$ standard solution in Table 3.42. Repeat the titration 3 times.

Data Recording and Processing

(1) Calculate the concentration of $AgNO_3$ standard solution, the mean value and the relative average deviation ($\bar{d}_r \leqslant 0.2\%$).

$$c_{AgNO_3} = \frac{m_{NaCl} \times \dfrac{25.00}{100.0}}{M_{NaCl} V_{AgNO_3}} \times 1000$$

(2) Calculate the mass fraction of chlorine in sample.

$$w_{Cl} = \frac{c_{AgNO_3} V_{AgNO_3}}{m_s \times \dfrac{25.00}{250.0}} \times \frac{M_{Cl}}{1000} \times 100\% \qquad (M_{NaCl} = 58.44 \text{ g/mol}, M_{AgNO_3} = 169.88 \text{ g/mol})$$

Table 3.41 Standardization of $AgNO_3$ standard solution

	1	2	3
V_{AgNO_3} (final)/mL			
V_{AgNO_3} (initial)/mL			
V_{AgNO_3} /mL			
c_{AgNO_3}/ (mol/L)			
$\bar{c}_{AgNO_3}$/ (mol/L)			
$\|d_i\|$			
$\bar{d}_r$ /%			

Table 3.42 Determination of chlorine content in sample

	1	2	3
m_s/g			
V_{AgNO_3} (final)/mL			
V_{AgNO_3} (initial)/mL			
V_{AgNO_3} /mL			
w_{Cl}/%			
$\bar{w}_{Cl}$/%			
$\|d_i\|$			
$\bar{d}_r$ /%			

Notes

(1) Shake vigorously during the titration process to release Cl^- adsorbed by AgCl precipitate in time to prevent the endpoint from being advanced.

(2) The titration should be under neutral or alkalescent condition. The optimal pH is 6.5~10.5. If there is

ammonium salt in the solution, the optimal pH is 6.5～7.2.

(3) The amount of indicator affects the experimental results. The appropriate concentration of the indicator is 5×10^{-3} mol/L.

(4) After the experiment, the burette with $AgNO_3$ solution should be rinsed with distilled water for 2～3 times, and then rinsed with tap water to avoid residual of AgCl precipitate. Waste water with silver should be recycled.

Questions

(1) What is the effect if the concentration of K_2CrO_4 indicator is too high or too low?

(2) Why should $AgNO_3$ solution be stored in a brown bottle and stored in the dark?

(3) Can Ag^+ be directly titrated using NaCl standard solution by Mohr method? Why?

Exp.28 Determination of Chlorine in Soluble Chloride by Vollhard Method

Objectives

(1) To learn the preparation and standardization of NH_4SCN standard solution.

(2) To master the principle of the determination of chlorine in chloride by back titration, using ammonium iron alum as the indicator.

Principles

In acid solution containing Cl^-, AgCl precipitate generates from excess $AgNO_3$ standard solution. Then the excess Ag^+ is titrated with NH_4SCN standard solution using ammonium iron alum as indicator. At the endpoint, the solution is red, which is the color of $[Fe(SCN)]^{2+}$ (the color of the solution is orange if the amount of the indicator is small). The reactions are as follows:

$$Ag^+ + Cl^- = AgCl\downarrow \text{(white)} \qquad K_{sp}=1.8\times10^{-10}$$

$$Ag^+ + SCN^- = AgSCN\downarrow \text{(white)} \qquad K_{sp}=1.0\times10^{-12}$$

$$Fe^{3+} + SCN^- = [Fe(SCN)]^{2+} \text{ (red)} \qquad K_1=138$$

If the concentration of $AgNO_3$ standard solution is c(mol/L); the consumed volume of NH_4SCN standard solution is V(mL) when performing standardization of NH_4SCN standard solution; the volume of $AgNO_3$ standard solution consumed in the determination of chlorine in sample is V_1 (mL); and the volume of NH_4SCN standard solution consumed in back titration is V_2 (mL), then the mass fraction of chlorine in sample can be calculated by the formula below:

$$c_{NH_4SCN}=\frac{25.00\times c_{AgNO_3}}{V}$$

$$w_{Cl}=\frac{35.45\times(c_{AgNO_3}\times V_1-c_{NH_4SCN}\times V_2)\times250}{1000\times m_s\times25.00}\times100\%$$

Materials

Equipment: acid burette (50 mL), volumetric flask (250 mL, 500 mL), pipette (25 mL), Erlenmeyer flasks (250 mL), beaker (100 mL), graduated cylinder (5 mL, 25 mL), analytical balance, brown reagent bottle.

Reagents: $AgNO_3$ solution (0.10 mol/L), NH_4SCN solution (0.10 mol/L), ammonium iron alum indicator (400 g/L), HNO_3 solution (8 mol/L), nitrobenzene, NaCl sample.

Procedures

(1) Preparation and standardization of $AgNO_3$ standard solution (0.10 mol/L).

See details in Exp. 27.

(2) Preparation and standardization of NH_4SCN standard solution (0.10 mol/L).

Weigh 3.8 g of NH_4SCN in a 100 mL beaker, dissolve with water, transfer to a 500 mL volumetric flask, dilute to the mark, mix well and store in a reagent bottle.

Pipette 25.00 mL of $AgNO_3$ standard solution in a 250 mL Erlenmeyer flask, add 5 mL of HNO_3 solution and 1 mL of ammonium iron alum indicator, then titrate with NH_4SCN standard solution until the solution is stable red. The Erlenmeyer flask should be swirled during the titration. Record the consumed volume of NH_4SCN standard solution V (mL) in Table 3.43. Repeat the titration 3 times.

(3) Determination of chlorine in the sample.

Accurately weigh 1.6 g of NaCl sample in a 100 mL beaker, dissolve with distilled water, transfer to a 250 mL volumetric flask, dilute to the mark with distilled water and mix well. Pipette 25.00 mL of the sample solution to a 250 mL Erlenmeyer flask, add 20 mL distilled water and 5 mL of HNO_3. Then, $AgNO_3$ standard solution was added using a burette until the excessive amount is 5～10 mL, and record the volume of $AgNO_3$ standard solution V_1 (mL). Add 2 mL of nitrobenzene, plug the Erlenmeyer flask with a stopper, swirl sharply for 30 s to separate the solution and nitrobenzene layer with AgCl precipitation. Add 1 mL of ammonium iron alum as indicator and titrate with NH_4SCN standard solution until the solution is light red. Record the consumed volume of NH_4SCN standard solution V_2 (mL) in Table 3.44. Repeat the titration 3 times.

Data Recording and Processing

Calculate the concentration of NH_4SCN standard solution and the chlorine content in the sample.

Table 3.43 Standardization of NH_4SCN standard solution

	1	2	3
m_{NH_4SCN} /g			
V (final)/mL			
V (initial)/mL			
V/mL			
c_{NH_4SCN} /(mol/L)			
$\bar{c}_{NH_4SCN}$ /(mol/L)			
$\lvert d_i \rvert$			
$\bar{d}_r$ /%			

Table 3.44 Determination of chlorine content in sample

	1	2	3
m_s /g			
V_1/mL			
V_2(final)/mL			
V_2(initial)/mL			
V_2/mL			
w_{Cl} /%			
$\overline{w}_{Cl}$ /%			
$\|d_i\|$			
$\overline{d}_r$ /%			

Notes

(1) The appropriate concentration of Fe^{3+} is 0.015 mol/L because its concentration affects the results.

(2) The concentration of H^+ should be controlled in the range of 0.1～1 mol/L. The Erlenmeyer flask should be swirled sharply during the titration.

(3) Nitrobenzene (toxic) or petroleum ether is added before titration to protect AgCl precipitate and avoid the consumption of titrant because AgCl can react with SCN^-.

Questions

(1) Why is HNO_3 used to acidify solution? Is it possible to use HCl or H_2SO_4? Why?

(2) Try to discuss the effect of acidity on the determination of halogen ion by Volhard method.

Exp.29 Determination of Barium in Barium Salt by Gravimetric Method

Objectives

(1) To understand the condition and procedure of precipitation method.

(2) To master the basic skills of gravimetric analysis.

Principles

Ba^{2+} can form a series of insoluble compounds, such as $BaCO_3$, $BaCrO_4$, $BaSO_4$, BaC_2O_4, etc. Among them, $BaSO_4$ meets the requirements of gravimetry for precipitate because of its poor solubility (dissolve 0.4 mg at 100℃ and only 0.25 mg at 25℃ in 100 mL solution), consistency between the composition and chemical formula, big molar mass and great chemical stability.

In order to obtain large particles and pure crystalline precipitate of $BaSO_4$, the sample is dissolved in water, acidified with dilute HCl solution, and heated to a slight boiling. An excess of dilute and hot H_2SO_4 solution is added dropwise under constant agitation to form a crystalline precipitate between Ba^{2+} and SO_4^{2-}. The precipitate is aged, filtered, washed, dried, charred, and burned, and weighed in the form of $BaSO_4$.

Materials

Equipment: beaker (250 mL), graduated cylinder (100 mL), funnel, quantitative filter paper, glass rod, crucible, Muffle furnace, watch glass, analytical balance.

Reagents: $BaCl_2$ sample, HCl solution (2 mol/L), H_2SO_4 solution (1 mol/L), HNO_3 solution (2 mol/L), $AgNO_3$ solution (0.1 mol/L).

Procedures

(1) Treatment of the sample.

Weigh 0.4～0.5 g of $BaCl_2 \cdot 2H_2O$ in a 250 mL beaker, add 100 mL of distilled water and 4 mL of HCl solution (2 mol/L). Dissolve the sample by stirring and heat to slight boiling.

(2) Precipitation.

Transfer 4 mL of H_2SO_4 (1 mol/L) to a 100 mL beaker, add 30 mL of distilled water, heat the solution to slight boiling. The hot and diluted H_2SO_4 was added drop by drop to the $BaCl_2$ solution with constant stirring.

(3) Aging.

Test for complete precipitation by adding 1～2 drops of H_2SO_4 solution (1 mol/L) to the clear supernatant liquid. Then cover the beaker with a watch glass. The solution is aged in a boiling water bath for half of an hour with stirring several times.

(4) Filtration and rinsing.

The slow quantitative filter paper is folded according to the angle of the funnel to make it fits well with the funnel. Wet with water and keep the water column inside the funnel neck. The funnel is placed on a funnel rack and a clean beaker is put under the funnel. The supernatant is carefully poured into the funnel along the glass rod, and then wash the precipitate 3～4 times with a pouring method, using 15～20 mL of the washing liquid each time (dilute 3 mL of 1 mol/L H_2SO_4 with distilled water to 200 mL). Finally transfer the precipitate quantitatively to the filter paper and wash the precipitate to the absence of Cl^- (tested by $AgNO_3$ solution).

(5) Carbonization, ashing and burning.

A clean crucible with stopper is burned at 800～850℃ until the weight is constant (m_1). The filter paper with precipitation is put into the crucible and dried by Muffle furnace. The crucible is then burned at 800～850℃ until the weight is constant. Two samples were determined in parallel. Record the mass m_1 (g) and m_2 (g) in Table 3.45.

Data Recording and Processing

Calculate the content of barium in sample.

$$w_{Ba} = \frac{\frac{M_{Ba}}{M_{BaSO_4}} \times (m_2 - m_1)}{m_{BaCl_2 \cdot 2H_2O}} \times 100\%$$

Table 3.45 Determination of Ba content

	1	2
m_s/g		
m_1/g		
m_2/g		
$(m_2 - m_1)$/g		
w_{Ba} /%		
$\overline{w}_{Ba}$ /%		
$\lvert d_i \rvert$		
$\overline{d}_r$ /%		

Notes

(1) The glass rod can not be taken out until the filtration and rinsing are finished during the whole process.

(2) During the precipitation process, the drop speed should not be too fast, and it should be continuously stirred to avoid excessive local concentration, and at the same time reduce the adsorption of impurities.

(3) When stirring, the glass rod should not touch the bottom wall of the beaker to avoid scratching the beaker, so that the precipitate adheres to the scratch of the beaker and is difficult to wash.

(4) During the aging process, the beaker should be placed obliquely on a small piece of wood to sink the precipitate and concentrate on one side of the beaker, to facilitating the separation and transfer of the precipitation.

Questions

(1) Why should $BaSO_4$ be precipitated in the medium of hot and diluted HCl solution? What is the effect if too much HCl solution is added?

(2) Why should $BaSO_4$ be precipitated in hot and dilute solution, while filtered after cooling? Why should it be aged after precipitation?

(3) Why should $BaSO_4$ be rinsed several times with small amount of rinsing solution?

(4) Why 0.4~0.5 g of $BaCl_2 \cdot 2H_2O$ sample is weighed? What is the effect if its mass is too much or little?

Exp.30 Determination of Barium in Barium Salts (Microwave Drying Gravimetric Method)

Objectives

To learn the measurement of barium in soluble salts via microwave oven drying constant weight of $BaSO_4$ precipitate.

Principles

The experimental principles and precipitation conditions are basically the same as those of

Exp. 29. When using a microwave oven to dry $BaSO_4$ precipitate to constant weight, the inside and outside of the sample are heated simultaneously, with no heat transfer process. Thus, the heating is rapid, uniform, and a higher temperature can be reached instantaneously. If high-boiling impurities such as H_2SO_4 are contained in the precipitate, it is difficult to decompose or volatilize impurities during the drying process of $BaSO_4$ precipitate by microwave heating. Therefore, in terms of precipitation conditions and washing operation, the Ba^{2+} containing test solution is further diluted, and the excess precipitant (H_2SO_4) is controlled within 20%～50%.

Materials

Equipment: beaker (100 mL, 250 mL), glass rod, microwave oven, circulating water vacuum pump (with suction bottle), G4 sand core crucible.

Reagents: H_2SO_4 solution (1 mol/L, 0.1 mol/L), HCl solution (2 mol/L), HNO_3 solution (2 mol/L), $AgNO_3$ solution (0.1 mol/L), $BaCl_2 \cdot 2H_2O$.

Procedures

(1) Preparation of precipitate.

See details in Exp. 29.

(2) Treatment of precipitate.

After the aging of newly prepared $BaSO_4$ precipitate, it was filtered and washed under reduced pressure with a G4 sand core crucible (m_1) which is already constant weighted in a microwave oven. Then, the crucible containing precipitate is dried in the microwave oven (10 min for the first time, and 4 min for second time). After drying, the crucible is transferred into a desiccator and cooled to room temperature (10～15 min), weighed, and repeated until constant weight is obtained (m_2) . Record the mass m_1 (g) and m_2 (g) in Table 3.46.

Data Recording and Processing

Calculate the amount of Ba in the sample.

Table 3.46 Determination of Ba content

	1	2
m_s /g		
m_1/g		
m_2/g		
(m_2-m_1)/g		
w_{Ba} /%		
$\bar{w}_{Ba}$ /%		
$\|d_i\|$		
$\bar{d}_r$ /%		

Notes

(1) Before using, the clean G4 sand core crucible is vacuumized for 2 min with a vacuum pump to remove moisture from the micropores of the glass sand plate for drying. It is placed in a microwave oven and dried at an output power of 500 W (medium and high temperature) for 10 min at first time and 4 min at second time. After each drying, it is placed in a desiccator for 10～15 min for cooling (leave a small gap when it is placed, and then cover it after about 30 s), and then quickly weigh it on an analytical balance. The difference between the masses weighed after drying must be less than 0.4 mg (constant weight).

(2) The use method and precautions for circulating water vacuum pump and microwave oven should be taught by the instructors or refer to the relevant instructions.

Questions

(1) What are the advantages of microwave heating technology in analytical chemistry (such as decomposition of samples and drying of samples)?

(2) How to carry out this experiment scientifically and reasonably to fully reflect the application characteristics of microwave heating technology in gravimetric analysis?

Exp.31 Determination of Potassium in Fertilizer

Objectives

(1) To grasp the preparation methods of fertilizer sample solutions.

(2) To study the gravimetric method for determining the potassium content using sodium tetraphenylborate as a precipitant.

Principles

The sodium tetraphenylborate reagent is added to the treated fertilizer sample to cause precipitation of potassium tetraphenylborate. The reaction is as follows:

$$Na[B(C_6H_5)_4] + K^+ \!=\!=\!= K[B(C_6H_5)_4] \downarrow + Na^+$$

The obtained $K[B(C_6H_5)_4]$ precipitate has the advantages of small solubility and good thermal stability. When the precipitate is formed, it is weighed and converted into the mass of K_2O after a series of treatments. The precipitation of potassium tetraphenylborate is carried out in an alkaline medium. The interference of ammonium ions can be masked by formaldehyde, while the interference of metal ions can be masked by disodium edetate.

Materials

Equipment: beaker (250 mL), volumetric flask (100 mL), pipette (25 mL), graduated cylinder (5 mL, 10 mL), watch glass, G4 sand core crucible, analytical balance.

Reagents: formaldehyde solution (25 g/L), disodium edetate solution (0.1 mol/L), phenolphthalein indicator (10 g/L), NaOH solution (20 g/L), potassium tetraphenylborate saturated solution (filtered until clear), sodium tetraphenylborate solution [0.1 mol/L, weigh 3.3 g of sodium tetraphenylborate, dissolve it in 100 mL of distilled water, add 1 g of $Al(OH)_3$, stir and leave it overnight, then filter repeatedly until clear], concentrated HCl, HNO_3 solution (1 mol/L).

Procedures

(1) Preparation of fertilizer sample solution.

Accurately weigh about 0.5 g of the inorganic fertilizer into a 250 mL beaker, add 20~30 mL of distilled water and 5~6 drops of concentrated HCl, cover the watch glass, and boil for 10 min at low temperature. After cooling, filter the residue and solution in the beaker to 100 mL volumetric flask, wash the inner wall of the beaker 5~6 times with hot distilled water, transfer the filtrate to the same volumetric flask, dilute to the mark with distilled water, and uniformly shake for use.

(2) Sample analysis.

Pipette 10~25 mL of the as-prepared fertilizer solution (depending on the potassium content in the sample) into a 250 mL beaker, add 5 mL of formaldehyde solution (25 g/L) and 10 mL of disodium edetate solution (0.1 mol/L). After mixing, add 2 drops of phenolphthalein indicator (10 g/L) and titrate with NaOH solution (20 g/L) until the solution is reddish. After that, heat the solution to 40℃, add 5 mL of sodium tetraphenylborate solution (0.1 mol/L) drop by drop, stir for 2~3 min, and stand for 30 min, then filter with constant weighted G4 sand core crucible (m_1). The crucible is then washed by potassium tetraphenylborate saturated solution for 2~3 times, and distilled water for 3~4 times (about 5 mL each time). After suction filtrating, the crucible is placed in a drying oven (or oven), dried at 120℃ for 1 h, then transferred into a desiccator and cooled to room temperature, weighed, and repeated until constant weight (m_2). Record the mass m_1 (g) and m_2 (g) in Table 3.47.

Data Recording and Processing

According to the weight of potassium tetraphenylborate precipitate, the mass fraction of K_2O in the fertilizer could be calculated.

Table 3.47　Determination of K_2O content

	1	2
m_s/g		
m_1/g		
m_2/g		
(m_2-m_1)/g		
w_{K_2O}/%		
$\overline{w}_{K_2O}$/%		
$\|d_i\|$		
$\overline{d}_r$/%		

Questions

(1) Why the NaOH solution is added before sodium tetraphenylborate solution?

(2) Why formaldehyde and disodium edetate solution is added during the measurement?

(3) Why the precipitate is washed with potassium tetraphenylborate saturated solution?

Exp.32 Determination of Nickel in Steel

Objectives

(1) To understand the principle and procedure for the determination of nickel by dimethylglyoxime nickel precipitation gravimetry.

(2) To master the basic operations of gravimetric analysis.

Principles

Dimethylglyoxime is a diprotic weak acid (represented by H_2D) with the molecular formula of $C_4H_8O_2N_2$ and the molar mass of 116.2 g/mol.

It has been proved that precipitation reaction occurs between H_2D and Ni^{2+} in ammonia solution:

$$Ni^{2+} + \begin{array}{l} H_3C - C = NOH \\ \qquad\quad | \\ H_3C - C = NOH \end{array} + 2NH_3 \cdot H_2O =\!=\!=$$

$$\begin{array}{ccccc} & O\cdots H-O & & & \\ & \uparrow & & & \\ H_3C - C = N & & N = C - CH_3 & & \\ & Ni & & \downarrow & + 2NH_4^{+} + 2H_2O \\ H_3C - C = N & & N = C - CH_3 & & \\ & | \qquad \downarrow & & & \\ & O - H\cdots O & & & \end{array}$$

The precipitate is filtered, washed, and dried at 120℃ to a constant weight to obtain the mass of the precipitate of dimethylglyoxime nickel, from which the mass fraction of Ni can be calculated. The reaction medium of the present method is an ammonia solution with pH of 8～9. Higher or lower acidity will increase the solubility of the precipitate. If the ammonia concentration is too high, an ammonia complex of Ni^{2+} is formed.

Dimethylglyoxime is a highly selective precipitant that only precipitates with Ni^{2+}, Pd^{2+} and Fe^{2+}. Water-soluble complex is formed between dimethylglyoxime and Co^{2+} or Cu^{2+}, and this process not only consumes H_2D, but also causes co-precipitation. It is preferred to carry out secondary precipitation or pre-separation when the content of Co^{2+} and Cu^{2+} is high. In addition, the existence of Fe^{3+}, Al^{3+}, Cr^{3+}, and Ti^{4+} interferes the measurement due to the formation of hydroxide precipitates in the ammonia solution. Therefore, citric acid or tartaric acid is added as complexing agent to form water-soluble complex before the addition of ammonia.

Materials

Equipment: analytical balance, beaker (500 mL), graduated cylinder (10 mL, 50 mL), watch glass, funnel, quantitative filter paper, G4 sand core crucible, vacuum pump, oven.

Reagents: mixed acid ($HCl : HNO_3 : H_2O=3 : 1 : 2$), tartaric acid or citric acid solution (500 g/L), dimethylglyoxime solution (10 g/L, in ethanol), ammonia (7 mol/L), HCl solution (6 mol/L), HNO_3 solution (2 mol/L), $AgNO_3$ solution (0.1 mol/L), ammonia-ammonium chloride washing solution(1 mL of ammonia water and 1 g of NH_4Cl in 100 mL of distilled water), a slightly ammoniated tartaric acid solution (20 g/L, pH 8～9), steel sample.

Procedures

Accurately weigh 2 parts of steel samples (containing 30～80 mg of Ni), place them in a 500 mL beaker, add 20～40 mL of mixed acid, cover the watch glass, dissolve at low temperature, boil to remove nitrogen oxides, and add 5～10 mL of tartaric acid solution (500 g/L, 10 mL per gram of sample) to the mixture with constant agitation. Then, ammonia (7 mol/L) is added dropwise until the pH of the solution is 8～9, at which time the solution turned blue-green. If insolubles are present, the precipitate should be filtered and washed several times with hot ammonia-ammonium chloride washing solution (the wash and the filtrate are combined). The filtrate is acidified with HCl (6 mol/L), diluted to about 300 mL with hot distilled water, heated to 70～80℃, and added with dimethylglyoxime ethanol solution (10 g/L) to precipitate Ni^{2+} (about 1 mL of 10 g/L dimethylglyoxime solution is needed with per milligram of Ni^{2+}) with constant stirring and finally 20～30 mL excess is added. The total amount of reagent added should not exceed 1/3 of the volume of the test solution to avoid increasing the dissolved amount of precipitation. Then, ammonia (7 mol/L) is added dropwise under constant stirring to adjust the pH of the solution to 8～9. Keep the mixture solution at 60～70℃ about 30～40 min, then filter it with a constant weight of G4 core crucible (m_1), wash the beaker and precipitate with slightly ammoniated tartaric acid solution (10 g/L) for 5～8 times. Wash the precipitate with warm distilled water until the solution without Cl^- (when testing Cl^-, the filtrate can be acidified with 2 mol/L HNO_3 solution and tested with 0.1 mol/L $AgNO_3$ solution). The sand core crucible with precipitation is placed in an oven at 130～150℃ for 1 h, then cooled to weigh, after the precipitation dried to weigh again until constant weight (m_2). Two samples were determined in parallel. Record the mass m_1 (g) and m_2 (g) in Table 3.48.

The sand core crucible is washed with diluted HCl solution after the experiment is completed.

Data Recording and Processing

The content of nickel in the sample was calculated based on the mass of the dimethylglyoxime nickel.

Table 3.48 Determination of Ni content

	1	2
m_s/g		
m_1/g		
m_2/g		
$(m_2 - m_1)$/g		
w_{Ni} /%		
$\overline{w}_{Ni}$ /%		
$\mid d_i \mid$		
$\overline{d}_r$ /%		

Questions

(1) What is the effect of adding HNO_3 when dissolving the sample?

(2) In order to obtain pure dimethylglyoxime nickel precipitation, what experimental conditions should be selected and controlled?

Chapter 4 Spectrophotometry and Common Separation Methods Experiment

Exp.33 Spectrophotometric Determination of Iron with Phenanthroline

Objectives

(1) To understand the components and the correct use of spectrometer.

(2) To learn the selection of experimental conditions for spectrophotometric analysis.

(3) To learn how to obtain the absorption spectrum, plot the standard curve and select maximum absorption wavelength.

Principles

The phenanthroline (phen) can react with Fe^{2+} to produce a stable red complex at pH 3~9. The $\lg K_f$ and molar absorption coefficient (ε) of the ferrous complex are 21.3 and 1.1×10^4 L/(mol·cm), respectively. The reaction is as follows:

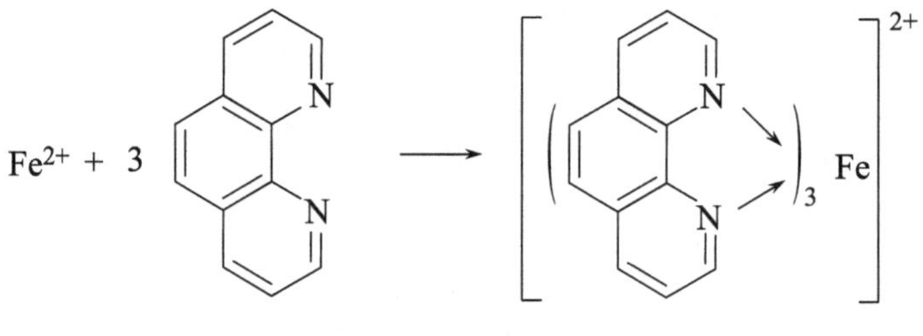

The maximum absorption of the red complex locates at 510 nm. The method is highly selective. The ions equivalent to 40-fold (such as Sn^{2+}, Al^{3+}, Ca^{2+}, Mg^{2+}, Zn^{2+}, SiO_3^{2-}), 20-fold [such as Cr^{3+}, Mn^{2+}, V(V), PO_4^{3-}] and 5-fold (such as Co^{2+}, Cu^{2+}) of the iron content have no obviously interference in the assay.

The color reaction is affected by various factors. Its conditions need to be determined by experiments.

Materials

Equipment: 721 (or 722) spectrophotometer, pipette (10 mL), volumetric flask (50 mL), cuvette (1 cm), pH meter.

Reagents: iron standard stock solution [0.1 mg/L, accurately weigh 0.7020 g of $NH_4Fe(SO_4)_2\cdot 6H_2O$ in a beaker, dissolve with 20 mL of H_2SO_4 (1∶1) and a small amount of water, and transfer to a 1 L volumetric flask. dilute the solution to the mark and mix well], iron standard solution (0.001 mol/L, obtained by the dilution of iron standard stock solution), phen solution (1.5 g/L, keep away from light, do not use when the color of the solution turns dark), hydroxylamine hydrochloride solution (100 g/L, freshly prepared), sodium acetate solution (1 mol/L), NaOH solution (0.1 mol/L), H_2SO_4 solution (1 : 1), iron-containing sample.

Procedures

(1) Optimization of experimental conditions.

(i) Selections of the absorption curve and detection wavelength.

Pipette 0 mL, 2 mL of iron standard solution into two 50 mL volumetric flasks, respectively. To each flask, add 1 mL of hydroxylamine hydrochloride solution, 2 mL of phen solution and 5 mL of NaAc. Fill each flask to the mark with distilled water, mix well and stand for 10 min. By employing the reagent blank (i.e. 0 mL of iron standard solution) as the reference solution, measure the absorbance with 1 cm cuvette in 10 nm increments in the range of 400～560 nm. Plot the graph of absorbance vs. the wavelength. Select the appropriate wavelength for determining Fe from the absorption curve, generally using the maximum absorption wavelength λ_{max}.

(ii) Selection of the amount of developer.

To each of the seven 50 mL volumetric flasks, add 2 mL of iron standard solution, 1 mL of hydroxylamine hydrochloride, and mix well. Then add 0.2 mL, 0.4 mL, 0.6 mL, 0.8 mL, 1.0 mL, 2.0 mL and 4.0 mL of the phen solution, respectively and supplement 5 mL of NaAc to each flask. Dilute to the mark with distilled water, mix well, and stand for 10 min. Measure the absorbance of each solution at the selected wavelength using 1 cm cuvette with distilled water as the reference solution. Plot the graph of absorbance vs. the volume of phen, and obtain the optimum amount of the developer for determination of iron.

(iii) Selection of the solution acidity.

To each of the seven 50 mL volumetric flasks, add 2 mL of iron standard solution, 1 mL of hydroxylamine hydrochloride and 2 mL of phen. Mix well, then add 0.0 mL, 2.0 mL, 5.0 mL, 10.0 mL, 15.0 mL, 20.0 mL and 30.0 mL of NaOH solution (0.1 mol/L), respectively. Dilute to the mark with water, mix well and stand for 10 min. Measure the absorbance of each solution at the selected wavelength using 1 cm cuvette with distilled water as the reference solution. Meanwhile, the pH of each solution was determined with a pH meter. Plot the graph of absorbance vs. pH of the solution to obtain the suitable acidity range.

(iv) Color development time.

In a 50 mL volumetric flask, add 2 mL of iron standard solution, 1 mL of hydroxylamine hydrochloride solution, and mix well. Add 2 mL of phen, 5 mL of NaAc, dilute to the mark with water, and mix well. Measure the absorbance at the selected wavelength using 1 cm cuvette immediately with distilled water as the reference solution. Measure the absorbance after the solution was stocked for 5 min, 10 min, 30 min, 60 min, 120 min. Plot the graph of absorbance vs.

color development time to obtain the appropriate time required for the complete reaction of iron with phenanthroline.

(v) Determination of the molar ratio of phenanthroline to iron.

Pipette 0.001 mol/L iron standard solution into each of the eight 50 mL volumetric flasks, add 1 mL of hydroxylamine hydrochloride solution and 5 mL of NaAc. Add 0.5 mL, 1.0 mL, 2.0 mL, 2.5 mL, 3.0 mL, 3.5 mL, 4.0 mL and 5.0 mL of phen, respectively. Dilute to the mark with water, mix well and stand for 10 min. Measure the absorbance of each solution at the selected wavelength using 1 cm cuvette with distilled water as the reference solution. Finally, plot the graph of absorbance vs. the concentration ratio of phenanthroline to iron (c_{Phen}/c_{Fe}) and determine the complexation ratio of Fe^{2+} ion to phenanthroline according to the intersection of the two extension lines on the curve.

(2) Determination of the iron content.

(i) Plotting the standard curve.

To each of the six 50 mL volumetric flasks, add 0.0 mL, 2.0 mL, 4.0 mL, 6.0 mL, 8.0 mL, 10.0 mL of iron standard solution (0.001 mol/L) with pipette. Add 1 mL of hydroxylamine hydrochloride, 2 mL of phen, 5 mL of NaAc, and mix well after adding each reagent. Then, fill each flask to the mark with water and stand for 10 min after mixing well. Measure the absorbance of each solution at the selected wavelength using 1 cm cuvette with a reagent blank (i.e. 0.0 mL of iron standard solution) as the reference solution. Plot the standard curve of absorbance vs. the concentration of iron, and calculate the molar absorption coefficient of the Fe^{2+}-phen complex.

(ii) Determination of iron in sample.

After the iron-containing sample solution was developed as step (i), the absorbance (A) was measured under the same conditions, and the iron content in the sample was determined from the standard curve.

Data Recording and Processing

Plot various experimental condition curves, standard curve and calculate the content of iron in the sample.

Questions

(1) Can the order for adding reagents be changed arbitrarily while making standard curve and performing the condition experiments? Why?

(2) What is the difference between the absorption curve and the standard curve?

(3) What are the roles of hydroxylamine hydrochloride and sodium acetate in this experiment?

(4) How to determine the content of the total iron and ferrous iron in water samples by spectrophotometry? Write out the basic steps.

Exp.34 Determination of NO_2^- in Food

Objectives

(1) To learn the operation of spectrophotometer.

(2) To master the method for the determination of NO_2^- in food.

Principles

As a food additive, nitrite can maintain the color and aroma of cured meat products and has certain antiseptic properties. But it also has a strong carcinogenic effect, and excessive intake is harmful to the human body.

In a weak acidic solution, the nitrite reacts with *p*-aminobenzenesulfonic acid to form a diazo dye, and the resulting diazo compound is coupled with naphthylethylenediamine hydrochloride to form a purple-red azo dye, which can be determined by spectrophotometric method. The relevant reactions are as follows:

$$NO_2^- + 2H^+ + H_2N-C_6H_4-SO_3H \longrightarrow N\equiv\overset{+}{N}-C_6H_4-SO_3H + 2H_2O$$

$$N\equiv\overset{+}{N}-C_6H_4-SO_3H + C_{10}H_7-NHCH_2CH_2NH_2\cdot HCl \longrightarrow$$

$$HO_3S-C_6H_4-N=N-C_{10}H_6-NHCH_2CH_2NH_2\cdot HCl$$

Materials

Equipment: 721 spectrophotometer, multi-purpose food shredder, volumetric flask (250 mL), filter paper, funnel, cuvette (2 cm).

Reagents: borax saturated solution (weigh 25 g of $Na_2B_4O_7\cdot 10H_2O$ and dissolve in 500 mL of hot water), zinc sulfate solution (1.0 mol/L, weigh 150 g of $ZnSO_4\cdot 7H_2O$ and dissolve in 500 mL of water), *p*-aminobenzenesulfonic acid solution (4 g/L, weigh 0.4 g of *p*-aminobenzenesulfonic acid and dissolve in 200 g/L hydrochloric acid to prepare a 100 mL solution, store the solution in the dark), naphthylethylenediamine hydrochloride solution (2 g/L, weigh 0.2 g of naphthylethylenediamine hydrochloride, dissolve in 100 mL of water and store the solution in the dark), $NaNO_2$ standard stock solution (accurately weigh 0.1000 g of analytical pure $NaNO_2$ dried for 24 h, dissolve in water and transfer to a 500 mL volumetric flask, dilute to the mark with water and mix well), $NaNO_2$ working solution (1 μg/mL, obtained by diluting the stock solution before use), activated carbon.

Procedures

(1) Sample pretreatment.

Weigh 5 g of the ground pickled meat sample into a 50 mL beaker, add 12.5 mL of borax saturated solution and stir the mixture well. Transfer the sample into a 250 mL volumetric flask by 150~200 mL of hot water (above 70℃) and put it in a boiling water bath for 15 min. Take out the flask and precipitate the protein by dropwise addition of 2.5 mL of $ZnSO_4$ solution with gently shaking. After cooling down to the room temperature, dilute to the mark with water, mix well and stand for 10 min. Remove the upper layer of fat, and filter the supernatant with filter paper or absorbent cotton. Discard the first 10 mL of the filtrate. The filtrate for the measurement should be colorless and transparent.

(2) Sample analysis.

(i) Plotting the standard curve.

Accurately pipette 0.0 mL, 0.4 mL, 0.8 mL, 1.2 mL, 1.6 mL, 2.0 mL of the $NaNO_2$ working solution (1 μg/mL) into six 50 mL volumetric flask respectively, add 30 mL of water, then add 2 mL of *p*-aminobenzenesulfonic acid solution and mix well. After standing for 3 min, add 1 mL of naphthylethylenediamine hydrochloride solution to each flask, dilute to the mark with water and mix well. After standing for 15 min, measure the absorbance of each test solution at wavelength of 540 nm using 2 cm cuvette with a blank reagent as the reference solution.

(ii) Determination of the sample.

Accurately transfer 40 mL of the pretreated sample filtrate into a 50 mL volumetric flask. The following procedure is the same as that in "Plotting the standard curve" (no dilution required). According to the measured absorbance, find the concentration of $NaNO_2$ from the standard curve.

Data Recording and Processing

Plot the standard curve with the amount of the $NaNO_2$ as the x-axis, and the absorbance as the y-axis. Calculate the mass fraction of $NaNO_2$ in the sample (expressed in mg/kg) based on the standard curve.

Notes

(1) Nitrite is easily oxidized to nitrate, thus the time and temperature of heating should be carefully controlled when processing the sample. In addition, the standard stock solution should not be stored for a long time.

(2) The content of nitrate in the sample is not included in the measurement by using this method.

Questions

(1) What are the characteristics of nitrite as a food additive? Can you find an alternative that is better than nitrite?

(2) Why do we discard the first 10 mL of filtrate when receiving the filtrate?

Exp.35 Determination of Available Phosphorus in Soil

Objectives

(1) To understand the principle and method of photometric determination of available phosphorus in soil.

(2) To be familiar with the operation of spectrophotometer.

Principles

The available phosphorus content in the soil refers to the amount of phosphorus that can be absorbed by the current crop. The most commonly used method for determining available phosphorus in soil is chemical method, in which the leaching agent (selected based on the nature of soil) is used to extract a portion of the available phosphorus from the soil. The available phosphorus in the form of iron phosphate and aluminum phosphate in acidic soil can be extracted with acidic ammonium fluoride to form ammonium fluoride aluminide and ammonium fluoride iron complex. A small amount of calcium ion forms calcium fluoride precipitate, and phosphate ion is extracted into the solution. The available phosphorus in calcareous soil is extracted with sodium bicarbonate solution. In the phosphorus-containing solution, ammonium molybdate is added, and under certain acidity condition, the phosphoric acid in the solution is complexed with molybdic acid to form the yellow phosphorus-molybdenum hybrid acid (phosphorus molybdenum yellow).

At an appropriate reagent concentration, a suitable reducing agent ($SnCl_2$ or ascorbic acid) is added to reduce a part of Mo (Ⅵ) in the phosphomolybdic acid to Mo (Ⅴ) to form phosphorus molybdenum blue (phosphorus molybdenum heteropoly blue, $H_3PO_4 \cdot 10MoO_3 \cdot Mo_2O_5$ or $H_3PO_4 \cdot 8MoO_3 \cdot 2Mo_2O_5$). Within a certain concentration range, the depth of blue is proportional to the phosphorus content, which is the basis of the molybdenum blue colorimetric method.

Materials

Equipment: 721 (or 722) spectrophotometer, volumetric flask (25 mL), plugged colorimetric tube (50 mL), pipette (25 mL), graduated cylinder (50 mL), shaker, funnel, filter paper.

Reagents: HCl solution (0.5 mol/L), NH_4F solution (1 mol/L), extractant (add 15 mL of 1 mol/L NH_4F solution and 25 mL of 0.5 mol/L HCl solution into 460 mL of distilled water to prepare a 0.03 mol/L NH_4F-0.025 mol/L HCl solution), H_3BO_3 solution (100 g/L), 15 g/L ammonium molybdate-3.5 mol/L hydrochloric acid solution (dissolve 15 g of ammonium molybdate in 300 mL of distilled water and heat to about 60℃. If there is precipitate, filter the solution. After the solution is cooled, slowly add 350 mL of 10 mol/L HCl solution and quickly stir with a glass rod. After cooling to the room temperature, dilute it to 1 L with distilled water, mix well and store in a brown bottle. It should not be stored for more than two months), stannous chloride solution (25 g/L, weigh 2.5 g of stannous chloride and dissolve in 10 mL of concentrated

HCl. After dissolution, add 90 mL of distilled water, mix well and put in brown bottle. This solution should be freshly prepared for use), phosphorus standard solution [50 μg/mL, accurately weigh 0.2195 g of KH_2PO_4 (AR, dried at 105℃), dissolve in 400 mL of distilled water, add 5 mL of concentrated H_2SO_4 to prevent mold, and transfer to 1 L volumetric flask. Dilute to the mark with distilled water and shake well. Accurately transfer 25.00 mL of the above solution into a 250 mL volumetric flask, dilute to the mark, and shake well to obtain 50 μg/mL phosphorus standard solution. This solution should not be stored for a long time].

Procedures

(1) Pretreatment of soil sample.

Weigh 1 g of soil sample (accurate to 0.01 g), put into a 50 mL plugged colorimetric tube, and add 20 mL of 0.03 mol/L NH_4F-0.025 mol/L HCl solution. Gently shake it by hands and then immediately place it on the shaker for 30 min. Filter with a non-phosphorus dry filter paper, and the filtrate is transferred into a 50 mL Erlenmeyer flask containing 15 drops of 100 g/L H_3BO_3 solution, and uniformly mix it (H_3BO_3 is added to prevent F^- from interfering with color development and etching the glass instrument).

(2) Determination of available phosphorus in soil.

Accurately remove 5～10 mL of the above soil filtrate into a 25 mL volumetric flask, add 5 mL of 15 g/L ammonium molybdate-hydrochloric acid solution with a pipette, shake well, add distilled water close to the mark, then add 3 drops of 25 g/L stannous chloride solution and dilute to the mark with distilled water and shake well. After 15 min of color development, the absorbance value is measured at 680 nm with 1 cm cuvette on a spectrophotometer with the reagent blank as the reference solution.

(3) Plotting the standard curves.

Accurately transfer 0.0 mL, 1.0 mL, 2.0 mL, 3.0 mL, 4.0 mL, 5.0 mL of 5 μg/mL phosphorus standard solution into six 25 mL volumetric flasks respectively, add 5～10 mL of 0.03 mol/L NH_4F-0.025 mol/L HCl solution (depending on the volume of filtrate), add 5 mL of ammonium molybdate-hydrochloric acid solution with pipette, add distilled water close to the mark, and add 3 drops of 25 g/L stannous chloride solution. Shake the flask until the solution is dark blue, then dilute the solution with distilled water to the mark, shake well, stand for 15 min, and then measure the absorbance of the solution under the same conditions as soil samples.

Data Recording and Processing

Plot the standard curve with the mass (μg) of phosphorus as the x-axis and the corresponding absorbance as the y-axis, and find out the phosphorus content in the soil sample from the standard curve.

Notes

The color of the phosphorus molybdenum salt reduced by stannous chloride is not stable enough, so the colorimetric time must be strictly controlled. Generally, the color is relatively stable within 15～20 min after color development, therefore, colorimetric measurement should be immediately operated when the solution is accurately placed for 15 min after color development, and complete the measurement in 5 min.

Questions

(1) Describe the basic principle of measuring phosphorus in this experiment.

(2) What is the influence when the stannous chloride solution has been stored for a long time?

Exp.36 Determination of Total Flavonoids in Bamboo Leaves by Spectrophotometry Coupled with Microwave-assisted Extraction Technique

Objectives

(1) To understand the principle and method of microwave-assisted extraction.

(2) To master the operation of microwave-assisted extraction for separation and extraction of real samples.

Principles

The microwave is an electromagnetic wave with a frequency between 300 and 300000 MHz, and a wavelength in the range of 1 mm to 1 m. Microwave-assisted extraction (MAE) is a new separation and extraction technology developed on the basis of traditional organic solvent extraction technology. By using appropriate solvents, according to the different physical and chemical properties of the compounds and the difference in microwave absorption capacity, microwave energy can be used to accelerate the separation of the target compound from the matrix and elute into the solvent, thereby improving the extraction efficiency of the target compound from the matrix.

Since the microwave radiation is a "body heating" process, the intracellular temperature rises rapidly when the natural product sample is extracted and separated, increasing the intracellular pressure sharply and leading to cell rupture, which is beneficial to the extraction agents getting into the cells during the reflux process. So that the effective ingredients contained in the cells are released quickly. Therefore, microwave-assisted extraction has the characteristics of fast, energy-saving, solvent-saving and environmentally friendly.

In recent years, it has been found that bamboo leaves contain a large number of active substances which are beneficial to human, including flavonoids, active polysaccharides and special amino acids, among which the phenolic compounds such as flavonoid glycosides are major. Modern scientific research shows that the flavonoids in bamboo leaves have excellent effects such as anti-free radical, anti-oxidation, anti-aging, anti-bacterial, anti-viral and protection of cardiovascular and cerebrovascular diseases. In this experiment, the flavonoids are extracted from bamboo leaves by microwave-assisted extraction, and the content of total flavonoids is determined by spectrophotometry.

Materials

Equipment: round bottom flask (50 mL), volumetric flask (50 mL), pipette (1 mL, 2 mL, 5 mL), microwave extractor, ultraviolet-visible spectrophotometer, centrifuge, analytical balance.

Reagents: anhydrous ethanol, rutin standard solution (300 μg/mL, accurately weigh 75 mg of rutin standard, dissolve and dilute to 250 mL with 60% ethanol, and shake well), $NaNO_2$ solution (50 g/L), $Al(NO_3)_3$ solution (100 g/L), NaOH solution (1 mol/L), bamboo leaf powder (cut the bamboo leaves, wash them with tap water and deionized water, then bake them in the oven for 7 h at 60℃. After the bamboo leaves are cooled to room temperature, they are chopped in a blender and passed through a 20 mesh sieve).

Procedures

(1) Microwave-assisted extraction of flavonoids from bamboo leaves.

Weight 1.00 g of bamboo leaf powder into a 50 mL round bottom flask, add 15 mL of 20% ethanol solution as extractant and shake well, put a magneton in the flask, set the speed to 500 r/min, and extract in a microwave extractor for 25 min at 75℃. After the extraction, the filtrate is filtered to remove chlorophyll and soluble lipids with petroleum ether, transferred to a 50 mL volumetric flask, and diluted to the mark with 20% ethanol solution.

(2) Plotting the standard curve.

Take 0.5 mL, 1.0 mL, 1.5 mL, 2.0 mL, 2.5 mL and 3.0 mL of rutin standard solution (300 μg/mL) into six 10 mL colorimetric tubes, respectively. Dilute to 5.0 mL with deionized water. Then add 0.50 mL of 50 g/L $NaNO_2$ solution, shake well and stand for 6 min; add 0.50 mL of 100 g/L $Al(NO_3)_3$ solution, shake well and stand for another 6 min; add 4.0 mL of 1 mol/L NaOH solution and dilute to 10 mL with deionized water, shake well, stand for 10 min and then centrifuge for 5 min. The solution is injected to 1 cm cuvette to measure the absorbance at 510 nm with the reagent blank as the reference solution. Record the data in Table 4.1.

(3) Sample analysis.

Pipette 5.00 mL of the microwave extract from bamboo leaves into a 10 mL colorimetric tube, determine the total flavonoids in the bamboo leaves by spectrophotometry according to the determination method of rutin standard solution, determine the total flavonoids content according to the standard curve (expressed as mass fraction). Record the data in Table 4.1.

Data Recording and Processing

Table 4.1 Absorbance of standard solution and determination of total flavonoids in sample

	The mass concentration of rutin/(mg/L)	Absorbance	The content of total flavonoids in sample/%
1			—
2			—

Continued

	The mass concentration of rutin/(mg/L)	Absorbance	The content of total flavonoids in sample/%
3			—
4			—
5			—
6			—
sample solution			

Questions

(1) What are the advantages of microwave-assisted extraction compared with conventional solvent extraction?

(2) What are the characteristics of microwave heating method? What are the main factors affecting microwave-assisted extraction?

Exp.37 Determination of Trace Lead in Environmental Water Samples by Extraction Separation-spectrophotometry

Objectives

(1) To understand the principle and method for the determination of lead in environmental water samples by dithizone extraction spectrophotometry.

(2) To master the basic operation of extraction separation.

Principles

Lead, an accumulated poison, is easily absorbed by the stomach and affects the metabolism of enzymes and cells through the blood. Excessive intake of lead will seriously affect human health, and its main toxic effects are anemia, neurological dysfunction and kidney damage etc. The "Sanitary Standard for Drinking Water" of China stipulates that the lead content in drinking water should not exceed 0.01 mg/L. Therefore, the content of lead in the environmental water sample is an important indicator of environmental monitoring.

The method of dithizone extraction spectrophotometry is one of the current national environmental standard monitoring methods for lead in water. The main principle of the method is that lead and dithizone can form a reddish dithizone chelate in a reducing medium of ammonia citrate-chloride-hydroxylamine hydrochloride at pH 8.5～9.5:

$$C_6H_5-N=N-\underset{\underset{SH}{|}}{C}=N-\overset{H}{N}-C_6H_5 + Pb^{2+} \longrightarrow$$

$$C_6H_5-\underset{H}{N}-N=\underset{\underset{S\ —\ Pb}{|}}{C}-N=\underset{\downarrow}{N}-C_6H_5 + H^+$$

The chelate can be extracted by an organic phase such as chloroform (or carbon tetrachloride), and has a maximum absorption wavelength at 510 nm with a molar absorption coefficient of 6.7×10^4 L/(mol·cm). Hydroxylamine hydrochloride is added to the sample to reduce Fe^{3+} and other oxidizing substances that may exist to prevent oxidation of dithizone; cyanide is added to mask Ag^+, Hg^{2+}, Cu^{2+}, Zn^{2+}, Cd^{2+}, Ni^{2+}, Co^{2+}, etc.; citrate is added to complex Al^{3+}, Cr^{3+}, Fe^{3+}, Ca^{2+}, Mg^{2+}, etc. to prevent these ions from hydrolyzing and precipitating in an alkaline solution. Dithizone extraction spectrophotometry has characteristics of high selectivity and sensitivity due to the extraction, separation and enrichment during the process. This method is suitable for the determination of trace lead in surface water and waste water.

Materials

Equipment: volumetric flask (250 mL), pipette (1 mL, 2 mL, 5 mL), spectrophotometer, separatory funnel (250 mL).

Reagents: lead standard solution [2.0 μg/mL, accurately weigh 0.1599 g of $Pb(NO_3)_2$ (purity ≥ 99.5%) and dissolve in about 200 mL of deionized water, add 10 mL of concentrated HNO_3, transfer to 1000 mL volumetric flask, dilute with distilled water to the mark. This solution contains 100.0 μg/mL lead. Pipette 10.00 mL of this solution into a 500 mL volumetric flask, dilute to the mark with distilled water, and shake well], dithizone stock solution (0.1 g/L, weigh 0.1000 g of dithizone and dissolve in 1000 mL chloroform, then store in a brown bottle at 0～4℃), dithizone working solution (0.04 g/L, pipette 100 mL of dithizone stock solution into a 250 mL volumetric flask and dilute to the mark with chloroform), dithizone special solution (250 mg of dithizone is dissolved in 250 mL of chloroform, this solution does not need to be purified, and is specially used for extracting purification reagent), reducing ammonia solution with citric acid-potassium cyanide [dissolve 100 g of diammonium hydrogen citrate, 5 g of anhydrous Na_2SO_3, 2.5 g of hydroxylamine hydrochloride, 10 g of KCN (highly toxic!) by distilled water, then dilute to 250 mL with distilled water and mix with 500 mL of ammonia].

Procedures

(1) Pretreatment of water sample.

Clean water (such as ground-water and clean surface water without suspended solids) can be directly measured. Otherwise, the pretreatment process is as follows.

(i) Turbid surface water: add 2.5 mL of concentrated HNO_3 to 250 mL of water sample, and slightly boil for 10 min on a hot plate. After cooling, filter it into a 250 mL volumetric flask with a quick filter paper, and the filter paper is washed several times with 0.03 mol/L HNO_3. Dilute to the mark.

(ii) Water sample contains more suspended matter and organic matter: add 10 mL of concentrated HNO_3 to 200 mL of water sample and boil to concentrate the solution to about 10 mL, slightly cool, add 10 mL of concentrated HNO_3 and 4 mL of concentrated $HClO_4$, and continue to digest to near dryness. After cooling, dissolve the residue with 0.03 mol/L HNO_3, cool and filter it into a 200 mL volumetric flask with fast filter paper, wash the filter paper with 0.03 mol/L HNO_3 and dilute to the mark.

(2) Plotting the standard curve.

Add 0 mL, 0.50 mL, 1.00 mL, 5.00 mL, 7.50 mL, 10.00 mL, 12.50 mL, 15.00 mL of lead standard solution to eight 250 mL separatory funnels respectively, dilute to 100 mL with deionized water, add 10 mL of 3 mol/L HNO_3 and 50 mL of reducing ammonia solution with citric acid-potassium cyanide, mix well. Then add 10.00 mL of dithizone working solution, shake it vigorously for 30 s, and stand to layering. Insert a lump of lead-free absorbent cotton into the neck of the separatory funnel, release the lower organic phase, discard the first 1～2 mL of the effluent, and then inject the organic phase into 1 cm cuvette. With chloroform as the reference solution, the absorbance is measured at 510 nm, and record the data in Table 4.2.

(3) Sample analysis.

Accurately measure the appropriate amount of the environmental water sample which is pretreated in step (1) in a 250 mL separatory funnel, dilute to 100 mL with deionized water, and the determination is carried out according to the measurement steps of standard curve. Record the data in Table 4.2.

Table 4.2 Calibration curve and determination of trace lead in sample

	Lead content/μg	Absorbance	The mass concentration of lead in sample/(μg/L)
1			—
2			—
3			—
4			—
5			—
6			—
7			—
8			—
water sample			

Data Recording and Processing

Plot the standard curve with the lead content (μg) as the x-axis, and the absorbance as the

y-axis. Calculate the lead content (μg) based on the standard curve. The mass concentration (μg/L) of lead in the environmental water sample is calculated from the volume of the water sample.

Questions

(1) Why is the extraction separation suitable for spectrophotometric determination of the trace lead in environmental water samples, but not applicable for mineral samples?

(2) Does the dithizone working solution need to be added accurately? Why?

(3) What is the purpose of water sample pretreatment?

Exp.38　Separation and Identification of Amino Acids by Paper Chromatography

Objectives

(1) To grasp the principle of paper chromatography for the separation and identification of amino acids.

(2) To grasp the operation technique of paper chromatography and the method of measuring the retention factor.

(3) To learn how to identify different components in an unknown sample based on the retention factors of the components.

Principles

Paper chromatography is a separation method using filter paper as a support, the moisture absorbed on the filter paper as the stationary phase, and the organic solvent as mobile phase. Capillary action can create flow of the mobile phase from the bottom to the top. The components in the sample will be distributed in the two phases continuously. Because of their different distribution coefficients, each molecule travels at a different speed along the piece of paper, forming the chromatographic points with different distances from the original spots and performing the separation. The retention factor (R_f) may be defined as the ratio of the distance traveled by the solute to the distance traveled by the solvent.

$$R_f = \frac{\text{distance traveled by a given analyte}}{\text{distance traveled in the same amount of time by the solvent front}} = \frac{a}{b}$$

Under certain conditions, the R_f value is a characteristic value of the substance, so it can be qualitatively analyzed according to R_f. There are many factors (such as the characteristics of solid phase and mobile phase, temperature, etc.) affecting the R_f value. Therefore, it is recommended to use the standard of each component for comparison in analysis.

The separation and identification of cystine, glycine and tyrosine (the R_f value increased in turn) is performed by paper chromatography. After the amino acid mixed sample is spotted on the filter paper, the sample is dissolved in the stationary phase, and the filter paper is immersed in the developing solvent (the mixture of n-butanol, glacial acetic acid and water). Due to the capillary action, the mobile phase moves up along the filter paper, and various amino acid components in the sample are continuously distributed in the stationary and the mobile phase. Due to their different partition coefficients, different solutes have different speeds moving with the mobile phase. Spots are formed at different distances from the original spots to achieve separation from each other. The amino acid itself is colorless, and the identified amino acid is usually developed with a ninhydrin coloring agent. That is, after chromatography, the developer ninhydrin needs to be sprayed on the paper, and the spots are red-purple. This method is much sensitive and can detect trace amino acids in micrograms.

Materials

Equipment: chromatography tank (150 mm × 300 mm, $\Phi \times h$), capillary, sprayer, medium speed chromatographic filter paper (cut into strips of 90 mm × 240 mm).

Reagents: developing solvent (*n*-butanol : glacial acetic acid : deionized water=4 : 1 : 2), amino acid standard solution (5 g/L of cystine, glycine and tyrosine aqueous solution, respectively), mixed solution of amino acid (mixed from the three amino acid standard solutions), ninhydrin solution (2 g/L, in *n*-butanol).

Procedures

(1) Spotting.

Draw a horizontal line about 3 cm from the edge of the strip and four equidistant points in pencil as the original spots on the line. The distance between the original spots is 2 cm. The three amino acid standard solutions and amino acid mixed test solution are sequentially spotted at the four original spots by capillaries, and the spot diameter is 2～2.5 mm. Put a cotton thread in the middle of 2 cm away from the other end of the filter paper and let it dry.

(2) Expand separation.

Pour 60 mL of developing solvent to the dry chromatography tank, and hang the sample strip on its lid. The lower end of the chromatography paper is immersed in the developer about 0.5 cm, but the original spots must leave the liquid surface. Cover the chromatography tank. When the solvent front edge rises to 2～3 cm from the upper end of the filter paper, the chromatography paper can be taken out and the solvent leading edge position can be drawn with a pencil.

(3) Coloration.

After the chromatography paper is dried in air, the ninhydrin solution is evenly sprayed on the filter paper with a sprayer. After drying, it is placed in an oven (about 90℃) for 3～5 min, and red-purple spots appear on the filter paper.

(4) Measuring.

Mark the range of the spot with a pencil, find the center of the spot, measure the distance *a* from the center of each spot to the starting line with a ruler, and measure the distance *b* from the solvent front to the starting line. Record the data in Table 4.3.

Data Recording and Processing

Calculate the values of R_f according to the values of a and b. Qualitatively identify the composition of the mixture by comparing the R_f of the amino acid in the mixed sample with the R_f of the standard amino acid.

Table 4.3　Analysis of amino acid mixture

<table>
<tr><th></th><th>a</th><th>b</th><th>R_f</th></tr>
<tr><td>cystine</td><td></td><td></td><td></td></tr>
<tr><td>glycine</td><td></td><td></td><td></td></tr>
<tr><td>tyrosine</td><td></td><td></td><td></td></tr>
<tr><td rowspan="3">mixture of amino acid</td><td></td><td></td><td></td></tr>
<tr><td></td><td></td><td></td></tr>
<tr><td></td><td></td><td></td></tr>
</table>

Notes

(1) During the experiment, the paper strip should be straight and in contact with the developing solvent, and the original spots should leave the liquid level.

(2) Note that the fingerprints contain a certain amount of amino acids (the skin can secrete amino acids). Do not touch the filter paper with your fingers directly. Use the tweezer to clamp the filter paper.

Questions

(1) What is the principle of separating amino acids by paper chromatography?

(2) What effect does it have on the experimental results if we use fingers to take the filter paper directly during the experiment?

(3) What will happen if the original spot is also immersed in the developing solvent?

(4) Why do standard sample often be used to identify unknown samples in paper chromatography?

Exp.39　Separation of Food Pigments by Paper Chromatography

Objectives

(1) To grasp the principle of paper chromatography for the separation of food pigments.

(2) To master the enrichment and determination method of the pigments in the sample.

Principles

The principle of paper chromatography is the same as Exp. 38. This experiment mainly focuses on the separation of synthetic pigments in beverages. After processing the sample, the artificial synthetic pigments are adsorbed by the polyamide under acidic condition, and separated from proteins, starches, fats and natural pigments, then the pigments are desorbed with a suitable desorption solution under alkaline conditions. Due to the different partition coefficients of different pigments, R_f is different and can be separated and identified.

Materials

Equipment: beaker (100 mL), analytical balance, chromatography tank (150 mm × 300 mm, $\Phi \times h$), sand core funnel (G2 or G3), chromatography paper (10 cm × 27.5 cm, $\omega \times h$), capillary (1 mm in diameter).

Reagents: pigment standard solution (5 g/L of carmine, tartrazine and sunset yellow), developing solvent (*n*-butanol : anhydrous ethanol : ammonium hydroxide=6 : 2 : 3), citric acid solution (200 g/L), polyamide powder (nylon 6 of 200 mesh, activated at 105℃ for 1 h previously), acetone (original), acetone-ammonia solution (the mixed solution of 90 mL of acetone and 100 mL of concentrated ammonium hydroxide).

Procedures

(1) Sample processing.

50 mL of orange juice without CO_2 is added into a 100 mL beaker, then the solution is adjusted to pH ≈ 4 with citric acid solution.

(2) Adsorption separation.

Add 0.5～1.0 g of polyamide into a 100 mL beaker, then get a homogenous paste with a small amount of distilled water and stir with the treated sample solution mentioned above at 70℃. The pigments in the sample solution will be completely adsorbed (if the polyamide is not enough, you can add more). The precipitate is separated by pump filtration in the sand core funnel and washed by distilled water (70℃) with stirring. Wash the precipitate twice with 20 mL of acetone to remove the grease. Then rinse the precipitate with 200 mL of distilled water (70℃) until the pH of filtrate is the same as original water. Note that stir fully during the process of washing.

Pigments are desorbed after several times of washing by acetone-ammonia solution (about 30 mL). Then the pigments are putted into a small beaker and adjusted to pH ≈ 6 by citric acid solution. Finally, the solution is concentrated to 5 mL by evaporation in water bath for spotting.

(i) Spotting.

Draw a horizontal line about 2.5 cm from the edge of the strip and four equidistant points named 1, 2, 3, 4 in pencil. Diffusion original spots with the diameter of 2 mm are separately spotted by capillaries with the standard solutions of carmine, lemon yellow and sunset yellow from No.1 to 3. Re-spot the original spot No.4 with the sample solution after drying with the hair dryer.

(ii) Expand separation.

After drying, the filter paper is hung on the cap with a hook, and then put it into the chromatographic tank with the developing solvent. Make sure that the filter paper is hung straight, the original points are 1 cm distant from the liquid level, keep the temperature at 20℃, and seal the chromatographic tank. The solvent is progressing up the paper. Remove the filter paper until it has raised almost 12 cm. After drying in the air, measure the distance of each spot traveled. Record the data in Table 4.4.

Data Recording and Processing

Calculate the R_f value for each spot based on the measured distance each spot traveled. The tested pigment and the standard pigment are the same pigment if they have the same R_f and color.

Table 4.4 Analysis of food pigments

	a	*b*	R_f
carmine			
tartrazine			
sunset yellow			
sample			

Notes

(1) Polyamide can adsorb acidic pigments under acidic conditions, because it is polymer compound. Make sure the acidic condition to prevent pigment from decomposing.

(2) The amide chains in the molecule can combine with the sulfonic acid group in the pigment by hydrogen bonds, so certain temperature and time are required for adsorption.

Questions

(1) What are the mobile phase and stationary phase in paper chromatography for separating the synthetic pigments?

(2) What should you pay attention to when washing polyamide? Why?

(3) Why should you adjust pH ≈ 4 in processing the sample?

Chapter 5 Comprehensive Design Experiments

The objective of this chapter for students are to further familiarize and solidify relevant knowledge and experimental operation skills such as weighing, pipetting and titration, develop ability of operating, analyzing and solving problems independently. Students should consult relevant references and make detailed protocol (including principles, preparation of reagents, preparation and standardization of standard solution, indicator, instruments, sampling amount, dissolution method of solid sample, specific analysis procedures and data processing, etc.) according to the experimental topics, and then carry out the experiment after teacher's review, finally finish the experimental report.

5.1 Acid-base Titration Design Experiments

Exp.40 Determination of SiO_2 by Potassium Fluosilicate Method

The alloy sample is decomposed by nitric acid and hydrofluoric acid to convert silicon into silicic acid, and then precipitates potassium fluorosilicate in the presence of large amount of potassium ions (the silicate sample is melted and decomposed by KOH and converted into a soluble silicate which forms a poorly soluble potassium fluorosilicate with KF in a strong acidic medium).

$$2K^{+} + SiO_3^{2-} + 6F^{-} + 6H^{+} = K_2SiF_6\downarrow + 3H_2O$$

The solubility of precipitation is relatively large, so KCl (s) is added during precipitation to reduce its solubility. Remove most of the free acids after filtration and washing, and the remaining free acids were neutralized with sodium hydroxide standard solution with phenolphthalein as indicator. The generated K_2SiF_6 precipitate is filtered off and hydrolyzed by adding boiling water, and the resulting HF can be titrated with sodium hydroxide standard solution. The reaction is as follows:

$$K_2SiF_6 + 3H_2O = 2KF + H_2SiO_3 + 4HF$$

The operation must be carried out in a plastic container because the generated HF has a corrosive effect on the glass.

References

Fu X S. 2015. Research of determination of silicon content in iron ally by potassium fluosilicate volumetric method. Fujian Analysis & Testing, 24(5): 41-42

Li H T, Dong Y Y, Liu X Y, et al. 2011. Determination of SiO_2 in siliceous refractories by potassium fluorosilicate-acid-base titration. Metallurgical Analysis, 31(2): 67-70

Exp.41　Determination of $NaHCO_3$ and Na_2CO_3 in Biscuits

Generally, Na_2CO_3 or $NaHCO_3$ is added to the biscuit during the preparation process to increase the crisp taste. To determine the content of each component in the sample, HCl standard solution can be used to titrate. According to the change of the pH value during titration, phenolphthalein and methyl orange are employed as indicators. Thus this method is often referred to as “double-indicator technique”.

In the double-indicator technique, phenolphthalein is generally used as the first indicator, followed by methyl orange. However it is not sharply change from reddish to colorless with phenolphthalein as indicator, the cresol red-thymol blue mixed indicator is often used. The discoloration range of cresol red is 6.7 (yellow)~8.4 (red), and that of thymol blue is 8.0 (yellow)~9.6 (blue). The discoloration point after mixing is 8.3, the acid color is yellow, and the alkali color is purple, and the mixed indicator is sensitive for color. The test solution is titrated with HCl standard solution from purple to pink, indicating the endpoint.

References

Huang W K. 1997. Food Inspection and Analysis. Beijing: China Light Industry Press, 592

Exp.42　Determination of Boron Trioxide of Slag

The method for preparing boric acid by decomposing boron-magnesium ore with sulfur dioxide has the advantages of simple and short process, high decomposition rate and recovery rate, which is a new process for producing boric acid. One of the main ways for the comprehensive utilization of slag is to produce boron-containing compound fertilizer after oxidation. Therefore, the determination of boron trioxide in slag is of practical significance.

Boric acid is a polyprotic and extremely weak acid ($K_a = 5.8 \times 10^{-10}$), thus it cannot be directly titrated with a base. However, borate can form the stable complex with glycerin, mannitol, etc., increasing the dissociation of boric acid in aqueous solution and converting boric acid into a medium strong acid. The reaction formula is as follows.

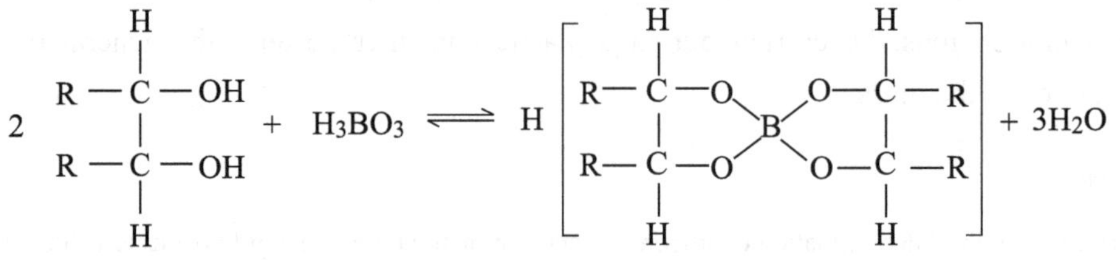

$$2\begin{array}{c} \text{H} \\ | \\ \text{R}-\text{C}-\text{OH} \\ | \\ \text{R}-\text{C}-\text{OH} \\ | \\ \text{H} \end{array} + H_3BO_3 \rightleftharpoons \text{H}\left[\begin{array}{ccccc} \text{H} & & & & \text{H} \\ | & & & & | \\ \text{R}-\text{C}-\text{O} & & & & \text{O}-\text{C}-\text{R} \\ | & & \text{B} & & | \\ \text{R}-\text{C}-\text{O} & & & & \text{O}-\text{C}-\text{R} \\ | & & & & | \\ \text{H} & & & & \text{H} \end{array}\right] + 3H_2O$$

The complex with the pK_a of 4.26 can be accurately titrated with NaOH standard solution. The titration reaction is as follows.

$$\mathrm{H}\left[\begin{array}{ccccc} & \mathrm{H} & & \mathrm{H} & \\ & | & & | & \\ \mathrm{R-} & \mathrm{C-O} & & \mathrm{O-C} & \mathrm{-R} \\ & | & \mathrm{B} & | & \\ \mathrm{R-} & \mathrm{C-O} & & \mathrm{O-C} & \mathrm{-R} \\ & | & & | & \\ & \mathrm{H} & & \mathrm{H} & \end{array}\right] + \mathrm{NaOH} = \mathrm{Na}\left[\begin{array}{ccccc} & \mathrm{H} & & \mathrm{H} & \\ & | & & | & \\ \mathrm{R-} & \mathrm{C-O} & & \mathrm{O-C} & \mathrm{-R} \\ & | & \mathrm{B} & | & \\ \mathrm{R-} & \mathrm{C-O} & & \mathrm{O-C} & \mathrm{-R} \\ & | & & | & \\ & \mathrm{H} & & \mathrm{H} & \end{array}\right] + H_2O$$

The reaction is carried out in equimolar, the pH of the solution at stoichiometric point is about 9.2, and phenolphthalein or thymol blue can be used as an indicator. Thereby, the measurement of the boron trioxide in the sample was performed.

References

Cheng X C. 2011. Comparison of strengthening reagents in boric acid determination. Chemical Engineer, 5: 60-61

Zhou X H. 2000. Determination of boron trioxide in slag. Physical Testing and Chemical Analysis (Part B: Chemical Analysis), 36(10): 473

5.2 Complexometric Titration Design Experiments

In complexometric titration, the first consideration is whether titration can be performed by controlling the acidity; secondly, the selection and application of masking agent is the key factor for the success of complexometric titration; for the selection of indicator, its acid-base property and the titration error caused by the complexation property should also be considered.

Exp.43 Determination of ZnO in Calamine Lotion

Calamine lotion is a commonly used drug for external use in clinical practice. Its main component is zinc oxide, and it also contains a small amount of iron oxide, magnesium oxide, calcium oxide, manganese oxide, etc. For the determination of calamine content, zinc oxide is used as the standard, and the content of zinc oxide can be determined by complexometric titration.

References

Wu X P. 2016. Identification and content determination of the calamine processed with the huanglian decoction. Clinical Journal of Chinese Medicine, 8(28): 31-32

Yang H, Shao Y, Chen C Y. 2013. A Study on the ZnO content determination method in calamine lotion. Journal of Nanjing Xiaozhuang University, 3: 59-62

Exp.44 Determination of Nickel and Magnesium in Nickel-magnesium Alloys

Nickel-magnesium alloy is an intermediate alloy (binary alloy) obtained by high-temperature melting of metallic nickel and metallic magnesium. It generally contains 70%～90% of nickel, 10%～30% of magnesium and a small amount of carbon, sulfur, iron, aluminum, copper, manganese and other elements. Presently, nickel-magnesium alloy can be used as a negative electrode material for nickel-hydrogen battery, and can be used as a spheroidizing agent in metallurgy to produce the working layer and core of the roll to change the state of graphite and increase the hardness of the roll. The main method for measuring high content of nickel and magnesium is titration with EDTA.

References

Li X Y, Guo C W, Luo X P. 2018. Determination of zinc content in alkaline zinc-nickel alloy plating bath by precipitation separation-titration. Plating and Finishing, 37(1): 29-31

Lu N P, Nian J Q, Zhang L F, et al. 2016. Determination of nickel and magnesium in nickel-magnesium alloy by EDTA titrimetry. Metallurgical Analysis, 36(1): 62-66

Exp.45　Determination of Each Component in Mixture of Mg^{2+} and EDTA

The first step is to check which component is excess at pH ≈ 10 using EBT as an indicator.

(1) If Mg^{2+} is excessive, pipette one test solution and titrate excess Mg^{2+} with EDTA. Take another test solution and adjust to pH = 5～6, titrate the total amount of EDTA with Zn^{2+} standard solution with XO as indicator.

(2) If the EDTA is excessive, pipette a test solution and adjust to pH = 5～6, determine the total amount of EDTA using Zn^{2+} standard solution with XO as indicator. Take another test solution and add NH_3-NH_4Cl buffer solution (pH≈10), then titrate excess EDTA using Zn^{2+} standard solution with EBT as indicator.

References

Experimental Center, College of Chemistry and Molecular Sciences, Wuhan University. 2013. Analytical Chemistry Experiment. 2nd ed. Wuhan: Wuhan University Press, 101-102

5.3　Redox Titration Design Experiments

Exp.46　Determination of Chromium in Alloys

In the sulfur-phosphorus mixed acid medium, using silver nitrate solution as a catalyst, the chromium (III) in the sample is oxidized to chromium (VI) with ammonium persulfate solution. The manganese sulfate solution is added, since chromium can be oxidized prior to manganese, when the Mn^{2+} is oxidized to a purple-red color, it can be judged that the chromium is completely oxidized and the manganese is simultaneously oxidized to permanganic acid. A small amount of hydrochloric acid solution is added and boiled to destroy the permanganic acid. Then titration can be carried out using ammonium ferrous sulfate standard solution with the phenyl substituted anthranilic acid solution as indicator.

When determining the chromium in the copper alloy, the sample is dissolved in nitric acid, and the perchloric acid smoke is taken to the mouth of the Erlenmeyer flask for a certain time. Then titration can be carried out using ammonium ferrous sulfate standard solution with the phenyl substituted anthranilic acid solution as indicator.

References

Li C Q, Wang L, Yang C X. 2016. Determination of chromium in copper alloy by perchloric acid treatment-ammonium ferrous sulfate titration. Metallurgical Analysis, 36(4): 71-74

Xu J Y, Mai L B, Chen X D. 2018. Determination of chromium in cobalt-chromium porcelain alloy by microwave digestion-ammonium ferrous sulfate titration. Metallurgical Analysis, 38(4): 74-78

Exp.47 Determination of Iodine in Iodized Salt

Salt is an indispensable condiment in our daily life. Iodine is a micronutrient element essential for human metabolism, growth and development. It is the main raw material for human body to synthesize thyroid hormone. The iodine in the iodized salt is mainly in the form of KIO_3, and KIO_3 can be reduced to I_2 by I^- in the acidic medium, and the resulting I_2 can be titrated with sodium thiosulfate standard solution using starch as indicator. The endpoint is observed when the blue color is just disappeared. Thereby the iodine content in the iodized salt could be obtained.

References

Li S D, Zhang Y, Zhang C H, et al. 2018. Study on the loss of the iodine in salt. Shanxi Chemical Industry, 38(2): 9-11

Exp.48 Determination of Ferrous Oxide in Iron-containing Steel Slag

The valence state and relative analysis of iron in steel slag is of great significance for the guidance of iron and steel smelting process. The separation and determination of metallic iron and ferrous iron in steel slag is usually performed by converting metallic iron to divalent iron into solution and then separates it by filtration. The ferrous iron in the sample is present in the insoluble residue. After acid dissolution, potassium dichromate titration is used to determine the ferrous oxide content.

References

Jia X, Deng H L, Tian X Z. 2017. Determination of ferrous oxide in steel slag containing iron by oxidimetry. Chemical Analysis and Meterage, 26(2): 89-91

5.4 Precipitation Titration Design Experiments

Exp.49 Determination of Sodium Chlorine in Food

Sodium chloride is present in variety of foods. It is important to the preservation time and the change of the internal character of food. GB/T 12457—2008 "Determination of Sodium Chloride in Foods" specifies methods for determining sodium chloride in foods, and indirect precipitation titration is one of them. After the test solution is acidified, an excess of silver nitrate solution is added, and excess silver nitrate is titrated with a potassium thiocyanate standard solution using ammonium ferric sulfate as indicator (Volhard method). The content of sodium chloride in the food is calculated according to the consumption of the potassium thiocyanate standard solution. It is also possible to use a direct precipitation titration method in which chloride ion is titrated with silver nitrate. When the chloride ion is completely precipitated by silver ions, the excess of silver ions and the potassium chromate indicator can form the brick red silver chromate precipitates, indicating the endpoint. The amount of sodium chloride in the food is calculated by the consumption of the silver standard solution.

References

Yan Z Y. 2016. Uncertainty evaluation for determination of sodium chloride in squid shreds by direct precipitation titrating method. Journal of Food Safety & Quality, 7(7): 2785-2789

Yu W, Han D, Zhu Q L. 2015. Proficiency testing results and analysis of sodium chloride in soy sauce. Journal of Anhui Agricultural Sciences, 43(18): 306-307, 309

Zou Y. 2010. Detection of sodium chloride in the soy sauce. Metrology and Measurement Technique, 1: 69-70

Exp.50 Determination of Chloride Ion in Complex Fertilizer

The content of chloride ion in complex fertilizer directly affects the growth of crops because the chlorine-free crops such as sweet potatoes, sugar beets, etc. are sensitive to chloride ions, and the dosage should be strictly controlled when fertilizing. The long-term use of complex fertilizer with high chloride ion content makes the soil acidification and salinization, which will affect the germination and emergence of seeds and inhibit the growth of crops. The national standard uses the Volhard method to determine the chloride ion content of the complex fertilizer. That is, the excess silver nitrate standard solution is firstly added to convert the chloride ion into a silver chloride precipitate, and then the precipitate is coated with dibutyl phthalate. Excess silver nitrate was back titrated with ammonium thiocyanate standard solution using ammonium ferric sulfate as indicator.

References

Du Y, Liu S J, Chen Y S. 2015. Study on detection method of chloride ion in organic fertilizers. Soil and Fertilizer

Sciences in China, 1: 111-114
Jiao L W. 2006. Determination of chloride content in fertilizer by Mohr method. Physical Testing and Chemical Analysis (Part B: Chemical Analysis), 42(3): 219-220
Shen Y, Cai W. 2017. Discussion on determination of chloride ion content in complex fertilizers. Journal of Zhejiang Agricultural Sciences, 58(10): 1783-1784
Zhou J L. 2005. Evaluation of the uncertainty of measurement for the determination of chloride ion content in compound fertilizer. Chemical Analysis and Meterage, 14(5): 7-10

Exp.51 Determination of Silver in Nylon Silver-plated Fiber

As a new type of textile fiber, silver-containing functional fiber has received extensive attention and application based on its functions such as deodorization, antibacterial, antistatic, radiation protection and medical care, etc. The functionality of the silver-containing functional fibers is mainly derived from the silver. Thus, the silver content is a key factor that directly affects its functionality. It is also an important factor for assessing the quality of silver-containing functional fibers and determining its price. A precipitation titration method can be employed for measuring the silver content of the silver-plated fiber. The silver-plated fiber is dissolved at a high temperature with concentrated sulfuric acid/concentrated nitric acid to change the silver into silver nitrate, and the substrate fiber is completely dissolved. A certain amount of concentrated hydrochloric acid is added to form a white precipitate of silver chloride. The silver chloride precipitate is separated by centrifugation, and then dried at a high temperature, and weighed the mass of the silver chloride. Thereby, calculate the silver content in the silver-plated fiber sample.

References

Guo K, Ma J W, Chen S J, et al. 2015. Determination of silver content of nylon silver-plated fiber by precipitation titration. Shandong Textile Technology, 2: 27-29

5.5 Spectrophotometry Design Experiments

Exp.52 Determination of Chromium and Manganese in Steel by Spectrophotometry

Both chromium and manganese are beneficial elements in steel, especially in alloy steels. The chromium and manganese are present not only as metal state in the solid solution of the steel, but also as carbides (CrC_2, Cr_5C_2, Mn_3C), silicides (Cr_3Si, MnSi, FeMnSi), oxides (Cr_2O_3, MnO_2), nitrides (CrN, Cr_2N) and sulfides (MnS), etc.

After the sample is dissolved by acid, Mn^{2+} and Cr^{3+} are formed. H_3PO_4 is added to mask the Fe^{3+}. Under acidic condition, an excess of $(NH_4)_2S_2O_8$ is added to oxidize Cr^{3+} and Mn^{2+} to $Cr_2O_7^{2-}$ and MnO_4^- using $AgNO_3$ as catalyst, respectively.

$$2Cr^{3+} + 3S_2O_8^{2-} + 7H_2O = Cr_2O_7^{2-} + 6SO_4^{2-} + 14H^+$$

$$2Mn^{2+} + 5S_2O_8^{2-} + 8H_2O = 2MnO_4^- + 10SO_4^{2-} + 16H^+$$

A strong absorption at 420～450 nm is observed for $Cr_2O_7^{2-}$, while absorption of MnO_4^- is very weak; MnO_4^- is strongly absorbed at 500～550 nm with double peaks, while absorption of $Cr_2O_7^{2-}$ is very weak. Measure the absorbance of the mixed solution at 440 nm and 545 nm, which are the maximum absorption wavelengths of the $Cr_2O_7^{2-}$ and MnO_4^-, respectively. According to the addition principle of the absorbance, the content of chromium and manganese in the test solution is determined by solving equations.

References

Central China Normal University, Northeast Normal University, Shanxi Normal University, et al. 2001. Analytical Chemistry Experiment. 3rd ed. Beijing: Higher Education Press

Experimental Center, College of Chemistry and Molecular Sciences, Wuhan University. 2013. Analytical Chemistry Experiment. 2nd ed. Wuhan: Wuhan University Press

Exp.53 Determination of Trace Chlorine in Air

The main sources of chlorine in the air are industrial waste gases such as chlor-alkali plants, the preparation of chlorine derivatives, and the synthesis of other chlorine-containing compounds. Sodium sulfite can be used to absorb chlorine in the air. Chlorine reacts with sodium sulfite to form chloride ions, excess sodium sulfite is oxidized to sulfate radical by hydrogen peroxide, and then excess hydrogen peroxide is decomposed by heating under alkaline conditions. After that, absorbance is performed by mercury thiocyanate method at 460 nm, which is the maximum absorption wavelength.

References

Gao X H, Ma Y L, Wu C S, et al. 2013. Determination of micro chlorine in air by absorptiometry. Chemistry & Bioengineering, 30(5): 85-87

Exp.54 Determination of Lead in Plant Leaves by Spectrophotometry

Lead is easily attached to plant foliage (leaves of trees and vegetables) due to environmental pollution. It is well known that lead is a kind of accumulated poisons, and excessive lead is very harmful to human body. The commonly used method for determining lead is dithizone color development with the potassium cyanide for masking. This method is sensitive and selective, but potassium cyanide has high toxicity and causes environmental pollution. Lead can also be determined with xylenol orange color development, in which phenanthroline is used as a masking agent. At pH 4.5～5.4, lead and xylenol orange form a stable 1 : 1 red complex, which has a maximum absorption at 580 nm, and the molar absorption coefficient is 1.55×10^4 L/(mol·cm).

References

Chen Z X, Fang F Y, Zheng B Y, et al. 2010. Photometric determination of adsorbed lead (Ⅱ) on leaves by spectrophotometry. Journal of Linyi Teachers College, 32(3): 105-109

Yang G J, Guo J, Du J H. 2000. Rapid determination of lead content in plant leaves by xylenol orange color development. Physical Testing and Chemical Analysis (Part B: Chemical Analysis), 36(9): 412-414

5.6 Comprehensive Experiments

Exp.55 Preparation and Composition Determination of Potassium Copper (Ⅱ) Dioxalate

Principles

Potassium copper (Ⅱ) dioxalate can be prepared by directly mixing copper sulfate with potassium oxalate, the reaction is as follows:

$$2K_2C_2O_4 + CuSO_4 \xlongequal{} K_2[Cu(C_2O_4)_2] + K_2SO_4$$

The solubility of potassium copper (Ⅱ) dioxalate in water is small, but an appropriate amount of ammonia water can be added to form copper ammonia ions to dissolve Cu^{2+} (pH about 10). It can also be dissolved by NH_3 -NH_4Cl solution.

The PAN indicator [1-(2-pyridylazo)-2-naphthol] belongs to the pyridine azo-based color developer, which is yellow in the pH range of 1.9～12.2. The complex of PAN indicator with copper ions is red, while the complex of copper ions with EDTA is blue. At the endpoint, the yellow color of the free PAN is mixed with the blue color of the EDTA-Cu complex to form the emerald green color.

Materials

Equipment: analytical balance, water circulation vacuum pump, beaker (100 mL, 400 mL, 600 mL, 800 mL), graduated cylinder (10 mL, 50 mL), glass rod, alcohol thermometer, watch glass, electric heating plate, Büchner funnel, suction bottle.

Reagents: $CuSO_4·5H_2O$ (s), $K_2C_2O_4·H_2O$ (s), $KMnO_4$ standard solution (0.02 mol/L), EDTA standard solution (0.02 mol/L), H_2SO_4 solution (3 mol/L), concentrated $NH_3·H_2O$, $Na_2C_2O_4$ (s), copper standard, NH_3-NH_4Cl buffer solution, PAN indicator (0.1%, in ethanol), HCl solution (6 mol/L).

Procedures

(1) Preparation of potassium copper (Ⅱ) dioxalate.

4 g of $CuSO_4·5H_2O$ is dissolved in 8 mL of water (90℃), and another 12 g of $K_2C_2O_4·H_2O$ is dissolved in 44 mL of water (90℃). The hot $K_2C_2O_4$ solution is quickly added to the $CuSO_4$

solution under intense stirring, and the precipitate is vacuum filtered after cooling to 10℃. 8 mL of cold water is used to wash the precipitate twice and the product was dried at 50℃.

(2) Determination of copper content.

Accurately weigh 0.17～0.19 g of the product, dissolve with 15 mL of NH_3-NH_4Cl buffer solution, add 50 mL of distilled water. Add 3 drops of PAN indicator, titrate with 0.02 mol/L EDTA standard solution (see details in Exp. 11) until the solution changes from light blue to emerald green.

(3) Determination of oxalate content.

Accurately weigh 0.21～0.23 g of the product, dissolve with 2 mL of concentrated ammonia, add 15 mL of H_2SO_4 solution (3 mol/L). There will be a light blue precipitate. Dilute the solution to 100 mL, and heat to 70～85℃ in water bath. Titrate the solution while it is still hot with 0.02 mol/L $KMnO_4$ standard solution (see details in Exp. 20) until the reddish color is appeared and not faded within 1 min. The precipitate gradually disappeared during the titration.

Based on the above analysis, calculate the content of Cu^{2+} and $C_2O_4^{2-}$ in the product, and derive the formula of the product.

References

Beijing Normal University, Northeast Normal University, Central China Normal University, et al. 2014. Inorganic Chemistry Experiment. 4th ed. Beijing: Higher Education Press

Exp.56 Preparation of Sodium Thiosulfate and Determination of Product Content

Principles

$Na_2S_2O_3 \cdot 5H_2O$ is a colorless and transparent crystal. It is soluble in water, and the solution is alkaline. It is immediately decomposed in the presence of acid. Sodium thiosulfate is used in analytical chemistry for the quantitative determination of iodine, as a dechlorination agent in the textile industry and the paper industry, as a fixer in the photographic industry, and as a first aid antidote in the pharmaceutical industry.

In this experiment, Na_2SO_3 and S are used to prepare sodium thiosulfate under boiling condition:

$$Na_2SO_3 + S \xlongequal{\triangle} Na_2S_2O_3$$

The sodium thiosulfate crystallized from the solution at room temperature is $Na_2S_2O_3 \cdot 5H_2O$. The content of sodium thiosulfate in the sample can be determined by $K_2Cr_2O_7$ as primary standard substance. $K_2Cr_2O_7$ reacts with KI to precipitate I_2:

$$Cr_2O_7^{2-} + 6I^- + 14H^+ \xlongequal{\quad} 2Cr^{3+} + 3I_2 + 7H_2O$$

The resulting I_2 can be titrated by $Na_2S_2O_3$ solution with starch solution as indicator until the blue color of the solution just disappears indicating the endpoint. Titration reaction is as follows:

$$I_2 + 2S_2O_3^{2-} \xlongequal{\quad} S_4O_6^{2-} + 2I^-$$

The concentration of $Na_2S_2O_3$ solution can be determined from the consumed volume of

$Na_2S_2O_3$ solution and the mass of $K_2Cr_2O_7$, and the content of $Na_2S_2O_3$ can be determined from the mass of the weighed $Na_2S_2O_3$.

Materials

Equipment: mortar, beaker (100 mL), funnel, evaporating dish, water bath kettle, analytical balance, volumetric flask (250 mL), pipette (25 mL), Erlenmeyer flask (750 mL).

Reagents: Na_2SO_3 (s), $K_2Cr_2O_7$ (s), $Na_2S_2O_3 \cdot 5H_2O$ (s), sulfur powder, HCl solution (6 mol/L), ethanol, KI solution (100 g/L), starch solution (5 g/L).

Procedures

(1) Preparation of $Na_2S_2O_3 \cdot 5H_2O$.

2 g of sulfur powder is weighed, ground and placed in a 100 mL beaker, and 1 mL of ethanol is added to make it wet. An additional 6 g of Na_2SO_3 and 30 mL of deionized water are added. Heat and stir until boiling, then use a small fire to heat, stir and keep slightly boiling for more than 40 min, until only a small amount of sulfur powder is suspended in the solution (the solution volume should be not less than 20 mL, appropriate water supplement is necessary in the reaction process). After hot filtration, the filtrate is transferred to an evaporating dish, heated in a water bath, and the filtrate is evaporated until some crystals are precipitated in the solution, and the crystals are cooled, that is, a large amount of crystals are precipitated (if no crystals are precipitated during a long cooling time, stir or add some $Na_2S_2O_3 \cdot 5H_2O$ crystal to promote crystal precipitation). Filter under reduced pressure and wash the crystals with a small amount of ethanol and drain. Dry at 40℃ (about 40 min), weigh and calculate the yield.

(2) Preparation of $Na_2S_2O_3$ standard solution.

Accurately weigh 6～7 g of $Na_2S_2O_3 \cdot 5H_2O$, dissolve in a small amount of freshly boiled and cooled deionized water, and then dilute to 250 mL with freshly boiled and cooled deionized water.

(3) Determination of $Na_2S_2O_3$ content in the product.

Pipette 25.00 mL of $K_2Cr_2O_7$ standard solution (0.017 mol/L) into a 250 mL iodine volumetric flask, add 5 mL of KI solution (100 g/L), 5 mL of HCl solution (6 mol/L), cover and shake, place in the dark for 5 min. Wait until the reaction is complete, dilute with 100 mL of distilled water, titrate with $Na_2S_2O_3$ standard solution until the solution is pale yellowish green, add 2～3 mL of starch solution (5 g/L), sequentially titrate the solution until the color turns from blue to bright green indicating the endpoint, record the consumed volume of $Na_2S_2O_3$ standard solution. Three measurements will be made in parallel.

Calculate the concentration of the $Na_2S_2O_3$ standard solution based on the recorded data. The percentage of $Na_2S_2O_3 \cdot 5H_2O$ was determined from the mass of the weighed $Na_2S_2O_3 \cdot 5H_2O$.

References

Beijing Normal University, Northeast Normal University, Central China Normal University, et al. 2014. Inorganic

Chemistry Experiment. 4th ed. Beijing: Higher Education Press
Writing Group of Inorganic and Analytical Chemistry Experiment, Nanjing University. 2015. Inorganic and Analytical Chemistry Experiment. 3rd ed. Beijing: Higher Education Press

Exp.57 Determination of Zinc Oxide and Boron Trioxide in Flame Retardants

Zinc borate ($2ZnO \cdot 3B_2O_3 \cdot 5H_2O$) is an inorganic additive flame retardant product. Its quality standard is usually expressed by the amount of zinc oxide and boron trioxide. The traditional determination method is to titrate the zinc ion with EDTA by controlling the pH in the range of 7～10 with eriochrome black T as indicator. Then sodium carbonate is employed to precipitate and separate the zinc ion, and the borate is converted into boric acid by using sulfuric acid. After neutralization with methyl red as indicator, mannitol and phenolphthalein indicator are added, the solution is titrated with a standard alkali solution to determine boric acid. Thereby the content of boron trioxide is obtained. It is also possible to directly determine the zinc content by complexometric titration to obtain the content of zinc oxide. Then, add an equimolar complexing titrant to react with zinc ion, and determine boric acid to obtain the content of boron trioxide.

References

Ren X H. 2002. Analysis and determination of flame retardant zinc borate. Shanxi Chemical Industry, 22(4): 33-34
Wang X D, Gong X G, Duan Q L. 2004. A novel approach to analyze boron trioxide content in zinc borate. China Elastomerics, 14(6): 61-63

Appendix

Appendix 1 Preparation of Commonly Used Buffer

Buffer	pK	pH	Preparation
aminoacetic acid-HCl	2.35 (pK_{a1})	2.3	weigh 150 g of aminoacetic acid, dissolve in 500 mL of water, add 80 mL of concentrated HCl, dilute to 1 L
H_3PO_4-citrate		2.5	after dissolving 113 g of $Na_2HPO_4 \cdot 12H_2O$ in 200 mL of water, add 387 g of citric acid, dissolve, dilute to 1 L after filtration
monochloroacetic acid-NaOH	2.86	2.8	dissolve 200 g of monochloroacetic acid in 200 mL of water, add 40 g of NaOH, dissolve, dilute to 1 L
potassium hydrogen phthalate-HCl	2.95 (pK_{a1})	2.9	dissolve 500 g of potassium hydrogen phthalate in 500 mL of water, add 80 mL of concentrated HCl, dilute to 1 L
formic acid-NaOH	3.67	3.7	dissolve 95 g of formic acid and 40 g of NaOH in 500 mL of water, dilute to 1 L
NaAc-HAc	4.74	4.7	dissolve 83 g of anhydrous NaAc in water, add 60 mL of glacial acetic acid, dilute to 1 L
Hexamethylene tetramine-HCl	5.15	5.4	dissolve 40 g of hexamethylene tetramine in 200 mL of water, add 10 mL of concentrated HCl, dilute to 1 L
Tris-HCl [tris hydroxyl methyl amino methane, $NH_2C(HOCH_2)_3$]	8.21	8.2	dissolve 25 g of Tris in water, add 8 mL of concentrated HCl, dilute to 1 L
NH_3-NH_4Cl	9.26	9.2	dissolve 54 g of NH_4Cl in water, add 63 mL of concentrated ammonia, dilute to 1 L

Appendix 2 Commonly Used Indicator

1. Acid-base Indicator

Indicator	pH range of color change	Color change	Preparation
methyl violet (first color change range)	0.13～0.5	yellow～green	1 g/L or 0.5 g/L aqueous solution
cresol red (first color change range)	0.2～1.8	red～yellow	dissolve 0.04 g of indicator in 100 mL of 50% ethanol
methyl violet (second color change range)	1.0～1.5	green～blue	1 g/L aqueous solution
thymol blue (first color change range)	1.2～2.8	red～yellow	dissolve 0.1 g of indicator in 100 mL of 20% ethanol
methyl violet (third color change range)	2.0～3.0	blue～purple	1 g/L aqueous solution

Continued

Indicator	pH range of color change	Color change	Preparation
methyl orange	3.1～4.4	red～yellow	1 g/L aqueous solution
bromophenol blue	3.0～4.6	yellow～blue	dissolve 0.1 g of indicator in 100 mL of 20% ethanol
congo red	3.0～5.2	blue purple～red	1 g/L aqueous solution
bromocresol green	3.8～5.4	yellow～blue	dissolve 0.1 g of indicator in 100 mL of 60% ethanol
methyl red	4.4～6.2	red～yellow	dissolve 0.1 g or 0.2 g of indicator in 100 mL of 60% ethanol
bromophenol red	5.0～6.8	yellow～red	dissolve 0.1 g or 0.04 g of indicator in 100 mL of 20% ethanol
bromothymol blue	6.0～7.6	yellow～blue	dissolve 0.05 g of indicator in 100 mL of 20% ethanol
neutral red	6.8～8.0	red～bright yellow	dissolve 0.1 g of indicator in 100 mL of 60% ethanol
phenol red	6.8～8.0	yellow～red	dissolve 0.1 g of indicator in 100 mL of 20% ethanol
cresol red	7.2～8.8	bright yellow～tea red	dissolve 0.1 g of indicator in 100 mL of 50% ethanol
thymol blue (second color change range)	8.0～9.6	yellow～blue	dissolve 0.1 g of indicator in 100 mL of 20% ethanol
phenolphthalein	8.2～10.0	colorless～purple red	dissolve 0.18 g of indicator in 100 mL of 60% ethanol
thymol	9.3～10.5	colorless～blue	dissolve 0.1 g of indicator in 100 mL of 90% ethanol

2. Acid-base Mixed Indicator

Composition of the indicator solution	Color point pH	Color		Notes
		acidic	basic	
three parts of 1 g/L bromocresol green ethanol solution one part of 2 g/L methyl red ethanol solution	5.1	purple-red	green	
one part of 2 g/L methyl red ethanol solution one part of 2 g/L methyl blue ethanol solution	5.4	red purple	green	pH 5.2 red purple pH 5.4 dark blue pH 5.6 green
one part of 1 g/L bromocresol green sodium salt solution one part of 1 g/L chlorophenol red sodium salt solution	6.1	yellow green	blue purple	pH 5.4 blue green pH 5.8 blue pH 6.2 blue purple
one part of 1 g/L neutral red ethanol solution one part of 1 g/L methyl blue ethanol solution	7.0	blue purple	green	pH 7.0 blue purple
one part of 1 g/L bromothymol blue sodium salt solution one part of 1 g/L phenol red sodium salt solution	7.5	yellow	green	pH 7.2 dark green pH 7.4 light purple pH 6.2 deep purple
one part of 1 g/L cresol red sodium salt solution three parts of 1 g/L thymol blue sodium salt solution	8.3	yellow	purple	pH 8.2 blue green pH 8.4 blue

3. Metallochromic Indicators

Indicator	Dissociation and color change	Preparation
eriochrome black T (EBT)	$H_2In^- \underset{}{\overset{pK_{a2}=6.3}{\rightleftharpoons}} HIn^{2-} \overset{pK_{a3}=11.5}{\rightleftharpoons} In^{3-}$ purple-red blue orange	5 g/L aqueous solution
xylenol orange (XO)	$H_3In^{4-} \overset{pK_a=6.3}{\rightleftharpoons} H_2In^{5-}$ yellow red	2 g/L aqueous solution
K-B indicator	$H_2In \overset{pK_{a1}=8}{\rightleftharpoons} HIn^- \overset{pK_{a2}=13}{\rightleftharpoons} In^{2-}$ red blue purple-red (acid chrome blue K)	dissolve 0.2 g of acid chrome blue K and 0.4 g of naphthol green B in 100 mL water
calcium indicator	$H_2In^- \overset{pK_{a2}=7.4}{\rightleftharpoons} HIn^{2-} \overset{pK_{a3}=13.5}{\rightleftharpoons} In^{3-}$ wine red blue wine red	5 g/L in ethanol
pyridine azo naphthol (PAN)	$H_2In \overset{pK_{a1}=1.9}{\rightleftharpoons} HIn^- \overset{pK_{a2}=12.2}{\rightleftharpoons} In^{2-}$ yellow green yellow light red	1 g/L in ethanol
Cu-PAN (CuY-PAN aqueous solution)	$CuY+PAN+M^{n+} = MY+ Cu\text{-}PAN$ light green colorless red	add 5 mL of HAc buffer solution (pH 5～6) and 1 drop of PAN indicator into 10 mL of 0.05 mol/L Cu^{2+} solution, heat to about 60℃. Titrat to green with EDTA to obtain 0.025 mol/L CuY solution. Transfer 2～3 mL into the test solution, then add a few drops of PAN solution
sulfosalicylic acid	$H_2In \overset{pK_{a1}=2.7}{\rightleftharpoons} HIn^- \overset{pK_{a2}=13.1}{\rightleftharpoons} In^{2-}$ colorless	10 g/L aqueous solution
calmagite	$H_2In^- \overset{pK_{a2}=8.1}{\rightleftharpoons} HIn^{2-} \overset{pK_{a3}=12.4}{\rightleftharpoons} In^{3-}$ red blue red orange	5 g/L aqueous solution

Note: The EBT, calcium indicator and K-B indicator are less stable in an aqueous solution, and they can be formulated into a solid powder having a mass ratio of indicator to NaCl of 1∶100 or 1∶200.

4. Redox Indicator

Indicator	$E^{\ominus}$ /V $[H^+] = 1$ mol/L	Color change		Preparation
		oxidation state	reduction state	
diphenylamine	0.76	purple	colorless	10 g/L in concentrated H_2SO_4
sodium diphenylamine sulfonate	0.85	purple red	colorless	5 g/L aqueous solution
N-*o*-phenylamino benzoic acid	1.08	purple red	red	0.1 g of indicator plus 20 mL of 50 g/L Na_2CO_3 solution, diluted to 100 mL with water
o-diphenanthrene-Fe (Ⅱ)	1.06	light blue	red	dissolve 1.485 g of *o*-diphenanthrene and 0.965 g of $FeSO_4$, dilute to 100 mL with water (0.025 mol/L)
5-nitro-phenanthroline-Fe (Ⅱ)	1.25	light blue	purple red	dissolve 1.608 g of 5-nitro-phenanthroline-Fe (Ⅱ) and 0.695 g of $FeSO_4$, dilute to 100 mL with water (0.025 mol/L)

5. Adsorption Indicator

Indicator	Preparation	For determination		
		measurable elements (titrant in brackets)	color change	detection conditions
fluorescent yellow	1% sodium salt aqueous solution	Cl^-, Br^-, I^-, SCN^- (Ag^+)	yellow green～pink	neutral or weak alkaline
dichlorofluorescein	1% sodium salt aqueous solution	Cl^-, Br^-, I^- (Ag^+)	yellow green～pink	pH = 4.4～7.2
tetrabromofluorescent yellow (blush)	1% sodium salt aqueous solution	Br^-, I^- (Ag^+)	orange red～red purple	pH = 1～2

Appendix 3 Concentrations and Densities of Common Acid-base Solutions

Reagent	Density/(g/mL)	w/%	c/(mol/L)
hydrochloric acid	1.18～1.19	36～38	11.6～12.4
nitric acid	1.39～1.40	65.0～68.0	14.4～15.2
sulfuric acid	1.83～1.84	95～98	17.8～18.4
phosphate	1.69	85	14.6
perchloric acid	1.68	70.0～72.0	11.7～12.0
glacial acetic acid	1.05	99.8 (GR), 99.5 (AR), 99.0 (CP)	17.4
hydrofluoric acid	1.13	40	22.5
hydrobromic acid	1.49	47.0	8.6
ammonia	0.88	25.0～28.0	13.3～14.8

Appendix 4 Common Primary Standard and Their Drying Conditions and Applications

Primary standard		Composition after drying	Dry condition, t/℃	Standardization substance
name	molecular formula			
sodium bicarbonate	$NaHCO_3$	Na_2CO_3	270～300	acid
sodium carbonate	$Na_2CO_3 \cdot 10H_2O$	Na_2CO_3	270～300	acid
borax	$Na_2B_4O_7 \cdot 10H_2O$	$Na_2B_4O_7 \cdot 10H_2O$	store in a desiccator containing NaCl and sucrose saturated solution	acid

Continued

Primary standard		Composition after drying	Dry condition, t/℃	Standardization substance
name	molecular formula			
potassium bicarbonate	$KHCO_3$	K_2CO_3	270～300	acid
oxalic acid	$H_2C_2O_4 \cdot 2H_2O$	$H_2C_2O_4 \cdot 2H_2O$	air drying at room temperature	base or $KMnO_4$
potassium hydrogen phthalate	$KHC_8H_4O_4$	$KHC_8H_4O_4$	110～120	base
potassium dichromate	$K_2Cr_2O_7$	$K_2Cr_2O_7$	140～150	reducing agent
potassium bromate	$KBrO_3$	$KBrO_3$	130	reducing agent
potassium iodate	KIO_3	KIO_3	130	reducing agent
copper	Cu	Cu	stored in a dryer at room temperature	reducing agent
arsenic trioxide	As_2O_3	As_2O_3	stored in a dryer at room temperature	oxidant
sodium oxalte	$Na_2C_2O_4$	$Na_2C_2O_4$	130	oxidant
calcium carbonate	$CaCO_3$	$CaCO_3$	110	EDTA
zinc	Zn	Zn	stored in a dryer at room temperature	EDTA
zinc oxide	ZnO	ZnO	900～1000	EDTA
sodium chloride	NaCl	NaCl	500～600	$AgNO_3$
potassium chloride	KCl	KCl	500～600	$AgNO_3$
silver nitrate	$AgNO_3$	$AgNO_3$	280～290	chloride
sulfamic acid	$HOSO_2NH_2$	$HOSO_2NH_2$	stored in vacuum H_2SO_4 dryer for 48 h	base
sodium fluoride	NaF	NaF	stored for 40～50 min at 500～550℃ in platinum crucible, cooling in H_2SO_4 dryer	

Appendix 5 Molecular Weight of Common Compounds

Molecular formula	Molecular weight	Molecular formula	Molecular weight	Molecular formula	Molecular weight
Ag_3AsO_4	462.52	CO_2	44.01	$CuCl_2 \cdot 2H_2O$	170.48
AgBr	187.77	$CO(NH_2)_2$	60.06	CuI	190.45
AgCl	143.32	CaO	56.08	$Cu(NO_3)_2$	187.56
AgCN	133.89	$CaCO_3$	100.09	$Cu(NO_3)_2 \cdot 3H_2O$	241.60
Ag_2CrO_4	331.73	CaC_2O_4	128.10	CuO	79.545
AgI	234.77	$CaCl_2$	110.99	Cu_2O	143.09
$AgNO_3$	169.87	$CaCl_2 \cdot 6H_2O$	219.08	CuS	95.61
AgSCN	165.95	$Ca(NO_3)_2 \cdot 4H_2O$	236.15	$CuSO_4$	159.60
$AlCl_3$	133.34	$Ca(OH)_2$	74.09	$CuSO_4 \cdot 5H_2O$	249.68
$AlCl_3 \cdot 6H_2O$	241.43	$Ca_3(PO_4)_2$	310.18	CuSCN	121.62
$Al(NO_3)_3$	213.00	$CaSO_4$	136.14	$FeCl_2$	126.75
$Al(NO_3)_3 \cdot 9H_2O$	375.13	$CdCO_3$	172.42	$FeCl_2 \cdot 4H_2O$	198.81
Al_2O_3	101.96	$CdCl_2$	183.32	$FeCl_3$	162.21
$Al(OH)_3$	78.00	CdS	144.47	$FeCl_3 \cdot 6H_2O$	270.30
$Al_2(SO_4)_3$	342.14	$Ce(SO_4)_2$	332.24	$FeNH_4(SO_4)_2 \cdot 12H_2O$	482.18
$Al_2(SO_4)_3 \cdot 18H_2O$	666.41	$Ce(SO_4)_2 \cdot 4H_2O$	404.30	$Fe(NO_3)_3$	241.86
As_2O_3	197.84	$CoCl_2$	129.84	$Fe(NO_3)_3 \cdot 9H_2O$	404.00
As_2O_5	229.84	$CoCl_2 \cdot 6H_2O$	237.93	FeO	71.846
As_2S_3	246.02	$Co(NO_3)_2$	182.94	Fe_2O_3	159.69
$BaCO_3$	197.34	$Co(NO_3)_2 \cdot 6H_2O$	291.03	Fe_3O_4	231.54
BaC_2O_4	225.35	CoS	90.99	$Fe(OH)_3$	106.87
$BaCl_2$	208.24	$CoSO_4$	154.99	FeS	87.91
$BaCl_2 \cdot 2H_2O$	244.27	$CoSO_4 \cdot 7H_2O$	281.10	Fe_2S_3	207.87
$BaCrO_4$	253.32	$CrCl_3$	158.35	$FeSO_4$	151.90
BaO	153.33	$CrCl_3 \cdot 6H_2O$	266.45	$FeSO_4 \cdot 7H_2O$	278.01
$Ba(OH)_2$	171.34	$Cr(NO_3)_3$	238.01	$FeSO_4 \cdot (NH_4)_2SO_4 \cdot 6H_2O$	392.13
$BaSO_4$	233.39	Cr_2O_3	151.99	H_3AsO_3	125.94
BiCl	315.34	CuCl	98.999	H_3AsO_4	141.94
BiOCl	260.43	$CuCl_2$	134.45	H_3BO_3	61.83

Continued

Molecular formula	Molecular weight	Molecular formula	Molecular weight	Molecular formula	Molecular weight
HBr	80.912	K_2SO_4	174.25	$Na_2S\cdot 9H_2O$	240.18
HCN	27.026	$MgCO_3$	84.314	$NaSCN$	81.07
$HCOOH$	46.026	MgC_2O_4	112.33	Na_2SO_3	126.04
CH_3COOH	60.052	$MgCl_2$	95.211	Na_2SO_4	142.04
H_2CO_3	62.025	$MgCl_2\cdot 6H_2O$	203.30	$Na_2S_2O_3$	158.10
$H_2C_2O_4$	90.035	$Mg(NO_3)_2\cdot 6H_2O$	256.41	$Na_2S_2O_3\cdot 5H_2O$	248.17
$H_2C_2O_4\cdot 2H_2O$	126.07	$MgNH_4PO_4$	137.32	$NiCl_2\cdot 6H_2O$	237.69
HCl	36.461	MgO	40.304	NiO	74.69
HF	20.006	$Mg(OH)_2$	58.32	$Ni(NO_3)_2\cdot 6H_2O$	290.79
HI	127.91	$Mg_2P_2O_7$	222.55	NiS	90.75
HIO_3	175.91	$MgSO_4\cdot 7H_2O$	246.47	$NiSO_4\cdot 7H_2O$	280.85
HNO_3	63.013	$MnCO_3$	114.95	P_2O_5	141.94
HNO_2	47.013	$MnCl_2\cdot 4H_2O$	197.91	$PbCO_3$	267.20
H_2O	18.015	$Mn(NO_3)_2\cdot 6H_2O$	287.04	PbC_2O_4	295.22
H_2O_2	34.015	MnO	70.937	$PbCl_2$	278.10
H_3PO_4	97.995	MnO_2	86.937	$PbCrO_4$	323.20
H_2S	34.08	MnS	87.00	$Pb(CH_3COO)_2$	325.30
H_2SO_3	82.07	$MnSO_4$	151.00	$Pb(CH_3COO)_2\cdot 3H_2O$	379.30
H_2SO_4	98.07	$MnSO_4\cdot 4H_2O$	223.06	PbI_2	461.00
$Hg(CN)_2$	252.63	NH_3	17.03	$Pb(NO_3)_2$	331.20
$HgCl_2$	271.50	CH_3COONH_4	77.083	PbO	223.20
Hg_2Cl_2	472.09	NH_4Cl	53.491	PbO_2	239.20
HgI_2	454.40	$(NH_4)_2CO_3$	96.086	$Pb_3(PO_4)_2$	811.54
$Hg_2(NO_3)_2$	525.19	$(NH_4)_2C_2O_4$	124.10	PbS	239.30
$Hg_2(NO_3)_2\cdot 2H_2O$	561.22	$(NH_4)_2C_2O_4\cdot H_2O$	142.11	$PbSO_4$	303.30
$Hg(NO_3)_2$	324.60	NH_4HCO_3	196.01	SO_2	64.06
HgO	216.59	$(NH_4)_2HPO_4$	132.06	SO_3	80.06
HgS	232.65	$(NH_4)_2MoO_4$	79.055	$SbCl_3$	228.11
$HgSO_4$	296.65	NH_4NO_3	80.043	$SbCl_5$	299.02
Hg_2SO_4	497.24	$(NH_4)_2S$	68.14	Sb_2O_3	291.50
$KAl(SO_4)_2\cdot 12H_2O$	474.38	NH_4SCN	76.12	Sb_2S_3	339.68
KBr	119.00	$(NH_4)_2SO_4$	132.13	SiF_4	104.08
$KBrO_3$	167.00	NH_4VO_3	116.98	SiO_2	60.084
KCl	74.551	NO	30.006	$SnCl_2$	189.62
$KClO_3$	122.55	NO_2	46.006	$SnCl_2\cdot 2H_2O$	225.65
$KClO_4$	138.55	Na_3AsO_3	191.89	$SnCl_4$	260.52
KCN	65.116	$Na_2B_4O_7$	201.22	$SnCl_4\cdot 5H_2O$	350.596
K_2CO_3	138.21	$Na_2B_4O_7\cdot 10H_2O$	381.37	SnO_2	150.71
K_2CrO_4	194.19	$NaBiO_3$	279.97	SnS	150.776
$K_2Cr_2O_7$	294.18	$NaCN$	49.007	$SrCO_3$	147.63
$K_3Fe(CN)_6$	329.25	Na_2CO_3	105.99	SrC_2O_4	175.64
$K_4Fe(CN)_6$	368.35	$Na_2CO_3\cdot 10H_2O$	286.14	$SrCrO_4$	203.61
$KFe(SO_4)_2\cdot 12H_2O$	503.24	$Na_2C_2O_4$	134.00	$Sr(NO_3)_2$	211.63
$KHC_2O_4\cdot H_2O$	146.14	CH_3COONa	82.034	$Sr(NO_3)_2\cdot 4H_2O$	283.69
$KHC_2O_4\cdot H_2C_2O_4\cdot 2H_2O$	254.19	$CH_3COONa\cdot 3H_2O$	136.08	$SrSO_4$	183.68
$KHC_4H_4O_6$	188.18	$NaCl$	58.443	$UO_2(CH_3COO)_2\cdot 2H_2O$	424.15
$KHSO_4$	136.16	$NaClO$	74.442	$ZnCO_3$	125.39
KI	166.00	$NaHCO_3$	84.007	ZnC_2O_4	153.40
KIO_3	214.00	$Na_2HPO_4\cdot 12H_2O$	358.14	$ZnCl_2$	136.29
$KIO_3\cdot HIO_3$	389.91	$Na_2H_2Y\cdot 2H_2O$	372.24	$Zn(CH_3COO)_2$	183.47
$KMnO_4$	158.03	$NaNO_2$	68.995	$Zn(CH_3COO)_2\cdot 2H_2O$	219.50
$KNaC_4H_4O_6\cdot 4H_2O$	282.22	$NaNO_3$	84.995	$Zn(NO_3)_2$	189.39
KNO_3	101.10	Na_2O	61.979	$Zn(NO_3)_2\cdot 6H_2O$	297.48
KNO_2	85.104	Na_2O_2	77.978	ZnO	81.38
K_2O	94.196	$NaOH$	39.997	ZnS	97.44
KOH	56.106	Na_3PO_4	163.94	$ZnSO_4$	161.44
$KSCN$	97.18	Na_2S	78.04	$ZnSO_4\cdot 7H_2O$	287.54

第 1 章　分析化学实验基本知识

1.1　分析化学实验的基本要求

分析化学实验是学习分析化学的一个重要环节，是化学和相关专业学生重要的必修基础课程。通过本课程的学习，学生可以加深对分析化学基本概念和基本原理的理解；掌握分析化学实验的基本操作和技能；培养严谨细致的工作作风和实事求是的科学态度。要学好这门课程，需做到以下几点：

(1) 实验前认真预习。结合实验教材和理论学习，领会实验原理，熟悉实验内容、方法、步骤及注意事项。写出预习报告，列出数据记录表格。

(2) 实验过程中，操作严格规范，仔细观察实验现象并及时、如实记录。学生应有专门的实验记录本，不得将数据随意记在单页纸或小纸片上。文字记录应整齐清洁，数据记录尽量采用表格形式按顺序有规律地表达。切忌弄虚作假、编造数据。

(3) 实验时应保持实验台整洁和实验室安静，实验完成后要及时清理实验台，仪器、药品摆放有序。

(4) 及时撰写实验报告。实验结束后，应认真整理、分析实验原始记录，并独立完成实验报告。实验报告一般包括实验名称、实验日期、实验目的、实验原理、仪器和试剂、实验内容和步骤、数据记录与处理、问题和讨论等。

1.2　实验室基本知识

1.2.1　实验室安全知识

分析化学实验中，经常使用有腐蚀性、易燃、易爆的化学试剂，易破损的玻璃仪器，某些精密仪器以及水、电、气等。为确保实验人员的人身安全和实验的正常进行，必须严格遵守以下实验室安全规则。

(1) 实验室内严禁饮食、吸烟。

(2) 实验完毕须洗手。水、电、气用完后应立即关闭，并在离开实验室前再次检查确认。

(3) 使用各种仪器时，要在教师讲解演示或阅读操作规程后，方可动手操作。

(4) 使用电器设备时，切忌用湿手开启开关，以免触电。

(5) 使用浓酸、浓碱及其他强烈腐蚀性试剂时，切勿溅在皮肤和衣服上。使用浓的 HNO_3、HCl、H_2SO_4、$HClO_4$、氨水时，均应在通风橱内进行。夏天开启浓氨水、HCl 时一定要先用自来水将其冷却，再打开瓶盖。

(6) 使用易燃的有机溶剂(如乙醚、丙酮、乙醇、苯等)时，一定要远离火焰和热源。用完立即盖紧瓶盖，置于阴凉处保存。低沸点的有机溶剂禁止在明火上直接加热，只能在水浴中加热。

(7) 热、浓的 $HClO_4$ 遇有机物常易发生爆炸。如果试样为有机物，应先加热浓硝酸，使其与有机物发生反应，有机物被破坏后再加入 $HClO_4$。$HClO_4$ 蒸发所产生的烟雾易在通风橱中凝聚。若经常使用 $HClO_4$，通风橱应定期用水冲洗，以免 $HClO_4$ 的凝聚物与尘埃、有机物作用，引起燃烧或爆炸。

(8) 使用汞盐、砷化物、氰化物等剧毒物品时应特别小心。特别注意氰化物不能接触酸，否则会产生剧毒的 HCN！含氰废液可先加 NaOH 调至 pH > 10，再加入过量的漂白粉，使 CN^- 氧化分解；或在碱性条件下加入过量的硫酸亚铁溶液，使 CN^- 转化为亚铁氰化物，然后作废液处理。严禁将含氰废液直接倒入下水道或废液缸中。

硫化氢气体有毒，有关硫化氢气体的操作一定要在通风橱中进行。

(9) 若发生烫伤，可在烫伤处抹上黄色的苦味酸溶液或烫伤软膏。严重者应立即送医院治疗。实验室若发生火灾，应根据起火的原因进行针对性灭火。汽油、乙醚等有机溶剂着火时，用沙土扑灭，此时绝对不能用水。导线或电器着火时，不能用水或 CO_2 灭火器，而应首先切断电源，用 CCl_4 灭火器灭火，并根据火情决定是否要向消防部门报告。

(10) 实验室应保持整齐、干净。不能将毛刷、抹布扔在水槽中。禁止将固体物、玻璃碎片等扔入水槽内，以免造成下水道堵塞。此类物质及废纸应放入废纸箱或实验室规定存放的地方。废酸、废碱应小心倒入废液缸，切勿倒入水槽内，以免腐蚀下水管。

1.2.2 分析化学实验用水

1. 分析化学实验用水规格

分析化学实验中最常用的溶剂和洗涤剂是纯水。应根据实验要求，选用不同规格的纯水。我国分析实验室用水的国家标准(GB/T 6682—2008)规定了实验室用水的技术指标、制备方法和检验方法。表 1.1 列出了实验室用水级别与主要指标。

表 1.1　实验室用水级别与主要指标

名称	一级	二级	三级
pH 范围(25℃)	—	—	5.0～7.5
电导率(25℃)/(mS/m)	≤ 0.01	≤ 0.10	≤ 0.50
可氧化物质含量(以 O 计)/(mg/L)	—	≤ 0.08	≤ 0.40
吸光度(254 nm，1 cm 光程)	≤ 0.001	≤ 0.01	—
蒸发残渣含量［(105 ± 2)℃］/(mg/L)	—	≤ 1.0	≤ 2.0
可溶性硅含量(以 SiO_2 计)/(mg/L)	≤ 0.01	≤ 0.02	—

在实际操作中，有些实验对纯水还有特殊的要求，可根据需要检验的有关项目，如氧、铁、氨含量等进行选择。

2. 纯水的制备

分析实验室用于溶解、稀释和配制溶液的水都必须经过纯化。分析实验的要求不同，对水质纯度的要求也不同。应根据不同要求，采用不同纯化方法制得纯水。一般实验室用的纯水有蒸馏水、去离子水、电导水、二次蒸馏水、无 CO_2 蒸馏水、无氨蒸馏水等。制备方法如下。

1) 蒸馏水

将自来水在蒸发装置上加热气化，然后将水蒸气冷凝得到蒸馏水。由于杂质离子一般不挥发，所以蒸馏水中所含杂质比自来水少得多，比较纯净，可达到三级水指标，但还是有少量的金属离子、CO_2 等杂质。

2) 去离子水

去离子水是将自来水或普通蒸馏水通过离子交换树脂柱后所得的水。其纯度比蒸馏水高，质量可达到二级或一级水指标，但离子交换树脂对非电解质及胶体物质无效，同时还会有微量的有机物从树脂中溶出。

3) 电导水

在第一套蒸馏器中装入蒸馏水，加入少量高锰酸钾固体，经蒸馏除去水中的有机物，制得二次蒸馏水。再将二次蒸馏水装入第二套蒸馏器中，加少量硫酸钡和硫酸氢钾固体，进行蒸馏。弃去馏头、馏后各 10 mL，收集中间馏分，即为电导水。

4) 二次石英亚沸蒸馏水

若想获得比较纯净的蒸馏水，可以进行重蒸馏。在准备重蒸馏的蒸馏水中加入适当的试剂，以抑制某些杂质的挥发。二次蒸馏水一般可达到二级指标。

5) 特殊用水

无氨蒸馏水：每升蒸馏水中加入 25 mL 5% NaOH 溶液后，再煮沸 1 h，然后检查铵离子；或每升蒸馏水中加入 2 mL 浓 H_2SO_4，再重蒸馏，即得无氨蒸馏水。

无 CO_2 蒸馏水：煮沸蒸馏水，直至体积为原来的 3/4 或 4/5 为止，隔离空气，冷却即得。无 CO_2 蒸馏水应储存于连接碱石灰吸收管的瓶中，其 pH 应为 7。

无氯蒸馏水：将蒸馏水在硬质玻璃蒸馏器中先煮沸，再进行蒸馏，收集中间馏分，即得无氯蒸馏水。

3. 水纯度的检查

电导率是纯水质量的主要指标。因此，可选用适宜的电导率仪测定纯水的电导率。也可根据具体实验要求测定纯水中某些杂质的含量。一般检查项目有以下几种。

1) 电阻率

25℃时，纯水的电阻率为 $(1.0\sim10)\times10^6\ \Omega\cdot\text{cm}$，超纯水为 $10\times10^6\ \Omega\cdot\text{cm}$。

2) pH

分析实验用纯水要求 pH 为 6～7。

3) Ca^{2+} 和 Mg^{2+}

取适量被检测的水，滴加 NH_3-NH_4Cl 缓冲液，pH 调至 10 左右，加入 1 滴铬黑 T 指示剂，不显红色。

根据具体实验要求，纯水检测项目有时还包括氯离子、铁离子、铜离子等。

1.2.3　玻璃仪器的洗涤

1. 洗涤方法

仪器洗涤是化学工作者进行化学实验的一项基本技术操作。定量分析使用仪器的洗涤干净程度直接关系到分析结果的精密度和准确度。

玻璃仪器的洗涤方法应根据实验要求、污物的性质和程度来选择。常用方法如下。

1) 初用玻璃仪器

新购买的玻璃仪器表面常附着碱性物质。可先用洗涤剂稀释液、肥皂水或去污粉等洗刷，再用自来水洗。然后浸泡在1%～2% HCl溶液中放置过夜或不少于4 h，再用自来水冲洗，最后用蒸馏水冲洗2～3次，在80～100℃烘箱内烘干备用。

2) 一般玻璃仪器

一般玻璃仪器如试管、烧杯、锥形瓶、量筒等，先用自来水洗刷至无污物，再选用大小合适的毛刷蘸取洗涤剂稀释液，仔细刷洗器皿内外。用自来水冲洗干净后，再用蒸馏水冲洗2～3次，烘干或倒置在清洁处。凡洗净的玻璃仪器，器壁上不应带有水珠，否则应再按上述方法重新洗涤。若发现内壁有难以去掉的污迹，应重新洗涤。

3) 量器

移液管、滴定管、容量瓶等量器使用后应立即浸泡于凉水中。实验完毕后用流水冲洗去附着的试剂、蛋白质等物质，晾干后在酸洗液中浸泡4～6 h(或放置过夜)，再用自来水充分冲洗，最后用蒸馏水冲洗2～4次，风干备用。

4) 其他容器

盛过各种有毒药品，特别是剧毒药品和放射性同位素等物质的容器必须经过专门处理，确保没有残余毒物存在方可进行清洗。

2. 常用的洗涤剂

1) 铬酸洗液

铬酸洗液是含有饱和 $K_2Cr_2O_7$ 的浓硫酸溶液。它具有很强的氧化性，适宜洗涤无机物、油污和部分有机物。配制方法是：将5 g粗 $K_2Cr_2O_7$ 研细，加10 mL热水溶解后，在不断搅拌下缓慢加入100 mL工业级浓 H_2SO_4，溶液呈暗红色，冷却后转入玻璃瓶中，备用。铬酸洗液可反复使用，当溶液呈绿色时，表明洗液已经失效，须重新配制。铬酸洗液腐蚀性很强，且六价铬对人体有害，使用时应注意安全。

2) $NaOH$-$KMnO_4$ 溶液

$NaOH$-$KMnO_4$ 溶液用于洗涤油污和某些有机物。配制方法是：将4 g $KMnO_4$ 溶于水中，再加入100 mL 100 g/L NaOH溶液。

3) 合成洗涤剂

合成洗涤剂用于洗涤油污和某些有机物，主要是洗衣粉、洗洁精等。

4) 盐酸-乙醇溶液

盐酸-乙醇溶液用于洗涤被有色物污染的比色皿、容量瓶和吸量管等。将化学纯盐酸和乙醇按1∶2的体积比混合即可。

5) 酸性草酸和盐酸羟胺溶液

酸性草酸和盐酸羟胺溶液适用于洗涤氧化性物质，如沾有 $KMnO_4$、MnO_2、Fe^{3+} 等的容器。配制方法是：取 10 g 草酸或 1 g 盐酸羟胺溶于 100 mL 1∶1 的盐酸溶液中即可。

1.2.4　化学试剂

1. 化学试剂的等级

化学试剂的纯度对实验结果影响很大，不同实验对化学试剂纯度的要求也不相同。因此，必须了解化学试剂的分类标准。表 1.2 是化学试剂等级。

表 1.2　化学试剂等级

等级	名称	符号	标签颜色	适用范围
一级试剂	优级纯	GR	绿色	精密分析实验
二级试剂	分析纯	AR	红色	一般分析实验
三级试剂	化学纯	CP 或 P	蓝色	一般化学实验
生化试剂	生物试剂	BR 或 CR	黄色	生物化学实验

其中，化学纯试剂杂质含量较少，用于一般化学实验。分析纯试剂杂质更少，用于分析测定实验。此外，还有一些特殊用途的试剂。例如，“光谱纯”试剂中的杂质低于光谱分析法的检测限，“色谱纯”试剂是在最高灵敏度时以 10^{-10} g 下无杂质峰表示的，“超纯试剂”用于痕量分析和一些科学研究工作。这些试剂的生产、储存和使用都有特殊的要求。

在分析工作中所选用试剂的级别要和所用的实验方法、用水、操作器皿等的等级相适应。在通常情况下，分析实验中所用的一般试剂可选用 AR 级试剂，并用蒸馏水或去离子水配制。在某些要求较高的实验中，若选用 GR 级试剂，则不宜使用普通蒸馏水或去离子水。在特殊情况下，当市售试剂纯度不能满足要求时，可自己动手精制。

2. 化学试剂的使用原则

在分析工作中选用试剂的纯度、级别要与所用的分析方法相当。在满足实验要求的前提下，要注意节约原则。

1) 不污染试剂

化学试剂不能用手接触。固体应用干净的药匙取用，取下的瓶盖应倒放在实验台上，用后立即盖好，防止污染和变质。化学试剂瓶盖不能弄混。

2) 节约试剂

实验操作中，化学试剂的用量应按规定取用。若未注明用量，则尽可能取少量。多余试剂不能倒回原试剂瓶，以免污染原试剂瓶中试剂。

3) 试剂专用

化学分析实验通常用分析纯试剂；仪器分析实验一般使用优级纯、分析纯或专用试剂。

3. 试剂的保管

在实验室中，试剂的保管是一项十分重要的工作。保管不好，可能造成有些试剂的变质失效，不仅浪费试剂，而且会使分析工作失败，甚至引起事故。通常情况下，化学试剂应保存在通风良好、干净、干燥的地方，防止水分、灰尘和其他物质污染。同时，特殊性质的试剂还应用不同方法保管。

(1) 易腐蚀玻璃的试剂(如氟化物、苛性碱等)应保存在塑料瓶或涂有石蜡的玻璃瓶中。

(2) 易氧化的试剂(如氯化亚锡、亚铁盐)和易风化或潮解的试剂(如 $AlCl_3$、无水 Na_2CO_3、NaOH 等)应用石蜡密封瓶口。

(3) 见光易分解的试剂(如 $KMnO_4$、$AgNO_3$ 等)应用棕色瓶盛装，并保存在暗处。

(4) 受热易分解的试剂、低沸点的液体和易挥发的试剂应保存在阴凉处。

(5) 吸水性强的试剂(如无水碳酸钠、苛性钠等)应严格密封。

(6) 相互发生化学反应的试剂(如氧化剂和还原剂)应分开存放。

(7) 剧毒试剂(如三氧化二砷、氯化汞等)必须特别妥善保管和安全使用。

第2章　定量分析基本操作

2.1　分 析 天 平

2.1.1　称量方法

根据不同的称量对象及称量要求，须采用相应的称量方法。常用的称量方法有以下三种。

1. 直接称量法

调定天平零点后，将称量物置于分析天平秤盘上，待天平达到平衡后，所得读数即为称量物的质量。要求称量物不易吸水，在空气中性质稳定，如金属、矿样、小烧杯等。

2. 固定质量称量法

这种方法用于称取某一固定质量的试剂。要求试样在空气中没有吸湿性，如金属、合金的粉末或小颗粒。先按直接称量法称取盛试样器皿的质量，然后按TAR键，再用小匙将试样逐步加到盛放试样的器皿中，直到天平达到平衡，显示数据与待称量的质量吻合。

3. 减量法

该法要求称出的试样质量在一定的范围内。常用于称取易吸湿、易氧化或易与CO_2反应的物质。先用洁净的小纸条套在称量瓶上，如图2.1(a)所示，从干燥器中取出装有试样的称量瓶放入分析天平秤盘上(注意不要让手指接触称量瓶任何部位)，按TAR键。取出称量瓶，放在盛试样容器的上方，用小纸片夹住瓶盖，打开瓶盖，将称量瓶倾斜，用瓶盖轻轻敲击称量瓶的上部，使试样慢慢落入容器中，如图2.1(b)所示。当倾出的试样接近所需的质量时，慢慢地将瓶竖起，再用瓶盖敲击瓶口上部，使黏在瓶口的试样落回瓶中，盖好瓶塞，再将称量瓶放回秤盘上称量。天平显示数值的绝对值即为所称试样质量。

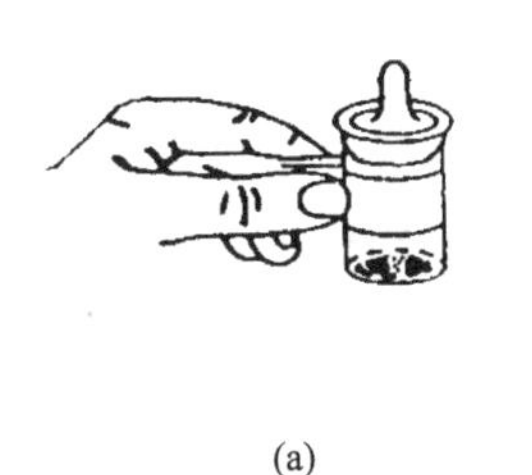

(a)　　(b)

图2.1　减量法

2.1.2　分析天平的使用方法

(1)称量前先将分析天平罩取下叠好，放在抽屉中。盛试样的容器放于天平左侧。

(2)检查分析天平的水平仪。如果水平仪的小气泡偏移，需调整水平调节脚，使小气泡位于水平仪圆圈内。

(3)开启天平，显示屏亮，同时天平进行自检，约 3 s 后显示天平的型号，然后进入称量模式，如 0.0000 g。新安装好的天平或长期未使用的天平使用前要进行校准，通常由 TAR(或清零)键、CAL 键及对应的校准砝码完成。

(4)按 TAR 键，显示为 0.0000 g 时，按所需的称量方法进行称量。称量时关好侧门，称量的数据应及时记录在专用记录本上，不得记在纸片或其他地方。在同一次实验中，应使用同一台分析天平，以减小称量误差。

(5)称量完毕，取出称量物，关好侧门，检查天平的零点。关闭分析天平，对其内外进行清洁，然后在登记簿上做好使用情况登记。再切断电源，最后罩上天平罩，将坐凳放回原处。

注意事项

(1)分析天平的载重不能超过它的最大负载。

(2)称量的物体必须与天平箱内的温度一致。不得把热的或冷的物体放在分析天平上称量。为了防潮，在天平箱内应放有吸湿用的干燥剂，如变色硅胶等。

(3)分析天平的前门不得随意打开。它主要供安装、调试和维修分析天平时使用。

(4)化学试剂和试样都不得直接放在秤盘上，应放在干净的表面皿、称量瓶或坩埚内。具有腐蚀性的气体或吸湿性物质必须放在称量瓶或其他适当的密闭容器中称量。

2.2　移液和定容操作

2.2.1　容量瓶

容量瓶是一种准确测量所盛装溶液体积的量入式量器。外形是细颈梨形的平底玻璃瓶，带有磨口玻璃塞或塑料塞。其颈上有一标线，一般指指定温度下，当溶液充满至弯月面下缘与标线相切时的溶液体积。常用的容量瓶规格有 10 mL、25 mL、50 mL、100 mL、250 mL、500 mL、1000 mL 等。

容量瓶主要用于配制准确浓度的标准溶液或定量地稀释溶液，使用容量瓶前应先检查是否漏水，标线位置离瓶口是否太近。如果漏水，则溶液不能准确配制；如果标线离瓶口太近，则不能充分混合溶液，故不宜使用。检漏时，加自来水至标线附近，盖好瓶塞，左手拿瓶颈标线以上部位，食指按住瓶塞，右手指尖托住瓶底边缘。倒立 2 min，如不漏水，将瓶直立，瓶塞转动 180°，再倒立 2 min，如不漏水，即可使用。使用容量瓶时，可用橡皮筋将瓶塞系在瓶颈上，以免被沾污或拿错后漏水。若瓶塞为平头的塑料塞，可将塞子倒置在台面上。容量瓶检漏完毕，使用前应洗涤干净。

溶液的配制：准确称取固体物质(基准试剂或被测试样)于小烧杯中，用一定量的水溶解。将溶液定量转移入容量瓶，转移溶液方法如图 2.2 所示。一手拿玻璃棒悬空伸入瓶口中，棒的下端靠在瓶颈内壁低于标线处；另一手拿烧杯，让烧杯嘴贴紧玻璃棒，慢慢倾斜烧杯，使溶液沿着玻璃棒流入容量瓶。转移完溶液后，将烧杯沿玻璃棒稍微向上提起，同时将烧杯直立，使附在玻璃棒和烧杯嘴之间的液滴回到烧杯中，再将玻璃棒放回烧杯中。用洗瓶以少量纯水吹洗烧杯和玻璃棒 3～4 次，每次的洗出液全部转入容量瓶中。然后用纯水稀释至容量瓶容积 2/3 处时，朝同一方向旋摇容量瓶使溶液混合(初步混匀)。此时勿倒转容量瓶！继续加水至标线以下约 1 cm 处，等待 1～2 min，使附在瓶颈内壁的溶液流下后，最后用滴管或洗瓶滴加纯水直至弯月面下缘与标线相切。最后，盖上干的瓶塞，左手捏住瓶颈标线以上部分，食指按住瓶塞，右手指尖托住瓶底边缘，将容量瓶倒转并摇动，再倒转过来，使气泡上升到顶。如此反复多次，使溶液充分混合均匀，如图 2.3 所示。

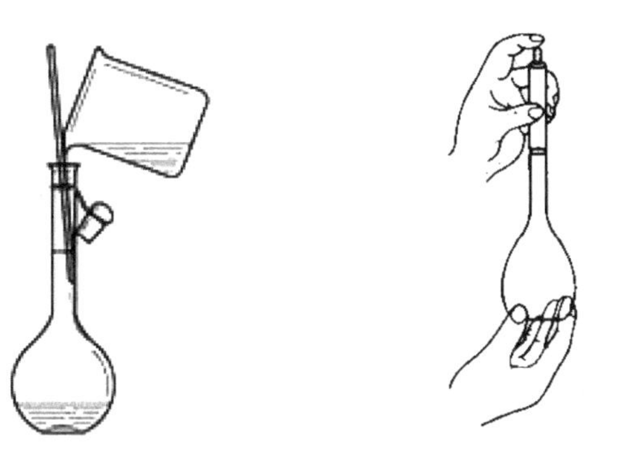

图 2.2　溶液定量转移　　　　图 2.3　溶液混匀

溶液的稀释：用移液管吸取一定体积的溶液于容量瓶中，按上述方法操作。

注意事项

(1) 热溶液应冷却至室温后再稀释，否则会造成体积误差。

(2) 需避光的溶液应以棕色容量瓶配制。

(3) 容量瓶不宜长期存放溶液。配好的溶液应转移到试剂瓶中保存，试剂瓶要先用配好的溶液润洗 2～3 次。

(4) 容量瓶使用完毕应立即用水冲洗干净。如长期不用，磨口处应洗净擦干，并用纸片将磨口隔开。

2.2.2　移液管

移液管是用于准确移取一定体积溶液的量出式量器。一种是细长而中间膨大的玻璃管，管颈上部有一环形标线，膨大部分标有容积和标定时的温度。在标明的温度下，吸取溶液至弯月面与管颈处标线相切，再让溶液按一定的方式自由流出，则流出溶液的体积就等于管上所标示的容积。另一种是带有刻度的玻璃管，一般用于量取较小体积的溶液。刻度有的刻到管尖，有的只刻到离管尖 1～2 cm 处。常用的移液管有 1 mL、2 mL、5 mL、10 mL、25 mL、50 mL 等各种规格(图 2.4)。

1. 移液管的洗涤

一般采用洗耳球吸取铬酸洗液洗涤移液管。也可放在洗液中浸泡，取出沥尽洗液后，用自来水冲洗，再用纯水润洗干净。润洗的水应从管尖放出。

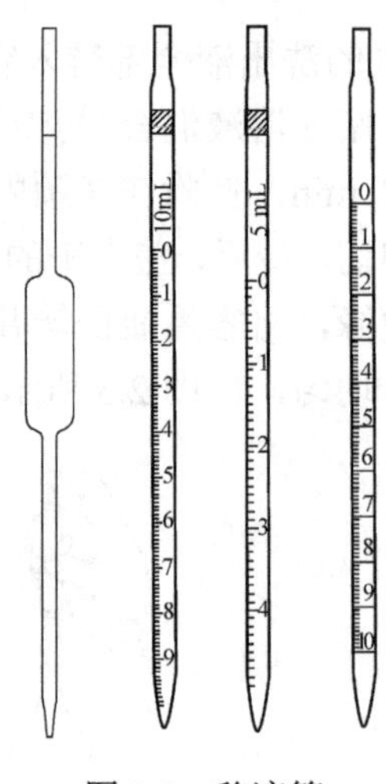

图 2.4　移液管

2. 移液管的使用

移取溶液前，用滤纸将移液管尖端内外的水吸尽，然后用待取溶液将移液管润洗 3 次。移取溶液时，一般用右手拇指和中指拿住管颈标线的上方，将润洗过的移液管管尖插入液面以下 1～2 cm，太深会使管外黏附的溶液过多，太浅易吸空。左手拿洗耳球，先把洗耳球内空气压出，然后将洗耳球的尖端接在移液管口，将管尖伸入溶液中吸取，如图 2.5(a)所示。移液管应随容器内液面的下降而下降。当管中液面上升到标线以上时，迅速移去洗耳球，立即用右手食指按住管口，将移液管提离液面，并将移液管原伸入溶液的部分，贴容器内壁转两圈，尽量除去管尖外壁黏附的溶液。然后将容器倾斜约 30°，竖直移液管，管尖紧贴容器内壁，略放松食指并用拇指和中指轻轻转动移液管，让溶液慢慢沿壁流出，直到溶液的弯月面下缘最低点与标线相切时，立即用手指压紧管口使溶液不再流出。将移液管移至盛接溶液

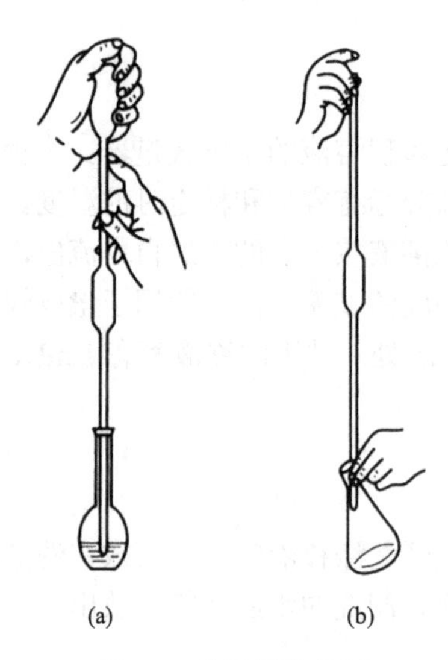

图 2.5　移液管的使用

的容器中，使管尖紧贴容器的内壁。移液管应呈垂直状态，盛接容器约成 30°倾斜。松开食指使溶液自由沿管壁流下，如图 2.5(b)所示。待溶液全部放完后，再等 15 s，取出移液管。刻度移液管放液时要使管内液面平稳下降，到所需的体积后即按紧食指，移去移液管。管上未标有“吹”字的，切勿把残留在管尖内的溶液吹入盛接的容器中，因为校正移液管时已经考虑了末端所保留溶液的体积。

2.3　滴定管及滴定分析基本操作

滴定管是滴定时用来准确测量流出标准溶液体积的量器，分为酸式滴定管和碱式滴定管(图 2.6)。酸式滴定管下端有玻璃旋塞，用来装酸性、中性和氧化性溶液，不宜盛碱性溶液(避免腐蚀磨口和旋塞)。碱式滴定管的下端连接一段乳胶管，管内有玻璃珠以控制溶液的流出速度，乳胶管下端再连一尖嘴玻璃管，主要用来装碱性和非氧化性溶液。凡是能与乳胶管反应的溶液，如 $KMnO_4$、I_2、硝酸银等，不得装在碱式滴定管中。目前市面上还有一种带聚四氟乙烯旋塞的通用型滴定管。这种滴定管可以克服上述酸式、碱式滴定管存在的旋塞易堵塞、乳胶管易老化及只宜装某些溶液的缺点，使用起来比较方便。

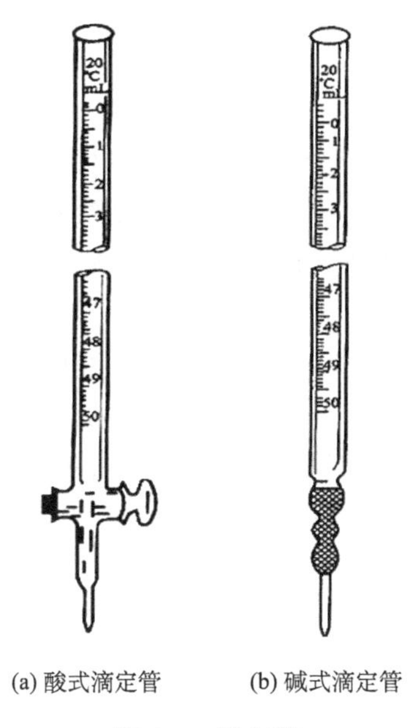

(a) 酸式滴定管　　(b) 碱式滴定管

图 2.6　滴定管

常用的滴定管标称容量为 50 mL 和 25 mL，还有标称容量为 10 mL、5 mL、2 mL、1 mL 的半微量或微量滴定管。标称容量为 50 mL 的滴定管最小刻度为 0.1 mL，读数可精准到 0.01 mL。

2.3.1　滴定管使用前的准备

使用酸式滴定管前，应检查旋塞转动是否灵活，然后试漏。试漏的方法为：先将旋塞关闭，在滴定管内充满水，将滴定管夹在滴定管夹上两分钟，检查是否漏水。将旋塞旋转 180°

再次确认不漏水。否则，应将旋塞取出，擦拭干净，重新涂上凡士林(气密封和润滑作用)后再使用。

涂凡士林的方法如图 2.7 所示。将滴定管中的水倒掉，平放在实验台上，抽出旋塞，用吸水纸将旋塞及旋塞槽的水擦干。用手指蘸少许凡士林，在旋塞的两头均匀地涂上薄薄一层，离旋塞孔近的两旁少涂一些，以免堵住旋塞孔。涂完凡士林后，将旋塞插入旋塞槽中，按紧。然后向同一方向转动旋塞，直至旋塞中油膜均匀透明。经上述处理后，旋塞应转动灵活，油脂层没有纹路。

图 2.7　涂凡士林的方法

碱式滴定管应选择大小合适的玻璃珠和乳胶管。玻璃珠过小会漏水或使用时上下滑动，过大则不便于放出液体。

滴定管使用前应洗涤干净，具体方法见 1.2.3 小节。

2.3.2　装入标准溶液

先用 5～10 mL 标准溶液润洗滴定管 2～3 次。操作时，两手平端滴定管，慢慢转动，使标准溶液布满全管，并使溶液从滴定管下端流尽，以除去管内残留水分。混匀后的标准溶液应直接倒入滴定管中，不得借助任何其他器皿，以免标准溶液浓度改变或造成污染。装好标准溶液后，应检查滴定管尖嘴内有无气泡，否则在滴定过程中，气泡溢出会影响溶液体积的准确测量。对于酸式滴定管，可迅速转动旋塞，使溶液快速冲出，将气泡带走。对于碱式滴定管，右手拿住滴定管上端，并使管身倾斜，左手捏挤玻璃珠周围乳胶管，并使尖嘴上翘，使溶液从尖嘴处喷出，即可排除气泡(图 2.8)。排除气泡后，装入标准溶液，使其在“0”刻度以上，再调节液面在 0.00 mL 处或稍下一点的位置，0.5～1 min 后，记录读数。

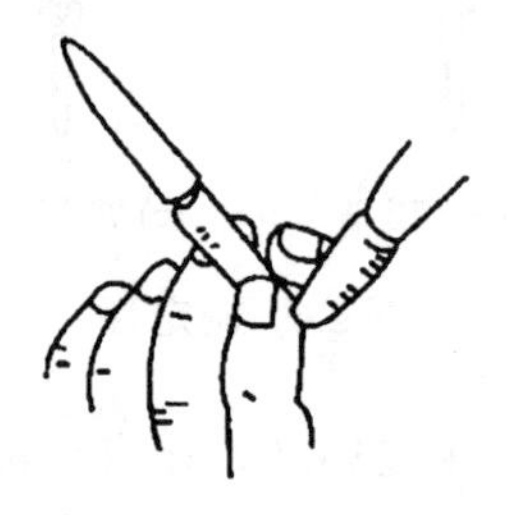

图 2.8　碱式滴定管中气泡的排除

2.3.3　滴定管的读数

滴定管读数不准确通常是造成滴定分析误差的主要原因之一。因此，读数时应遵循下列规则：

(1) 装满溶液或放出溶液后，须等 1～2 min，使附着在内壁的溶液流下来后再进行读数。如果放出溶液的速度较慢或临近终点，可等待 0.5～1 min 后读数。每次读数前都要检查管尖是否有气泡，是否悬有液滴。

(2) 读数时应将滴定管从滴定管架上取下，拇指和食指捏在滴定管液面的上方，使滴定管保持垂直状态。

(3) 对于无色或浅色溶液，读数时视线应与溶液弯月面下缘最低点处液面呈水平。注意初始读数与最终读数应采用同一标准。

(4) 使用蓝带滴定管时，液面呈现三角交叉点，此时应读取交叉点处的刻度。

(5) 每次滴定前应将液面调节在 0.00 mL 处或稍下一点的位置，以减少误差。

(6) 读数必须读到小数点后第 2 位，要求准确到 0.01 mL。

(7) 为了读数准确，可采用读数卡，这种方法有助于初学者练习读数。读数时，将读数卡放在滴定管背后，读取黑色部分与溶液弯月面下缘相切的刻度。读数方法应保持一致，即均使用读数卡，或均不使用读数卡。

2.3.4　滴定操作

滴定时，应将滴定管垂直地夹在滴定管架上，滴定台应呈白色，否则应放一块白瓷板作背景，以便观察滴定过程中溶液颜色的变化。滴定最好在锥形瓶中进行，必要时也可以在烧杯中进行。滴定操作如图 2.9 所示。

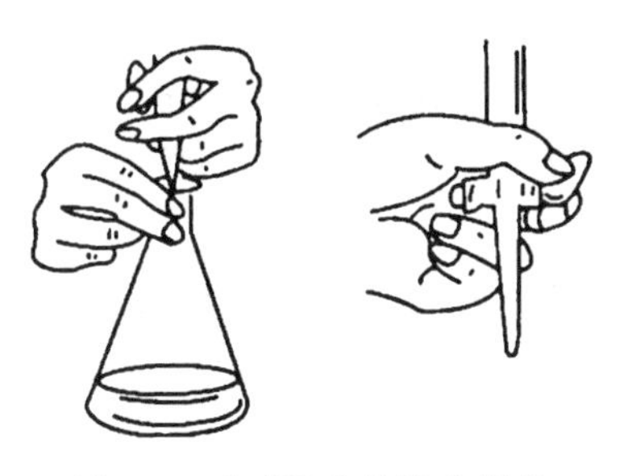

图 2.9　酸式滴定管滴定操作

酸式滴定管：用左手控制旋塞，拇指在前，食指和中指在后，手略微弯曲，轻轻向内扣住旋塞。转动旋塞时要注意勿使手心顶着旋塞，以防旋塞松动，造成溶液渗漏。右手握住锥形瓶，使滴定管尖稍伸进瓶口为宜。边滴边振荡，使瓶内溶液混合均匀，反应及时、完全。滴定时，左手不要离开旋塞，并注意观察溶液颜色的变化。临近终点时，滴定速度要减慢，滴一滴，摇几下，并用洗瓶吹入少量纯水洗锥形瓶内壁。

碱式滴定管：使用时左手拇指在前，食指在后，其余三指夹住出口玻璃管。用左手拇指与食指的指尖捏挤玻璃珠周围右侧上方的乳胶管，使乳胶管与玻璃珠之间形成一小缝隙，溶液即可流出。应当注意，不要用力捏玻璃珠，也不要使玻璃珠上下移动。停止加液时，应先松开拇指和食指，然后再松开其余三指。

2.4　重量分析法的基本操作

重量分析的基本操作包括样品溶解、沉淀、过滤、洗涤、烘干(或灼烧)、称量等。任何过程的操作正确与否都会影响最后的分析结果，因此每步操作均要细心准确。

2.4.1　样品溶解

样品溶解方法有两种，一种是用蒸馏水或酸溶解，另一种是高温熔融后再用溶剂溶解。

2.4.2　沉淀

对晶形沉淀的沉淀条件应做到“五字原则”，即稀、热、慢、搅、陈。

稀：沉淀的溶液配制要适当稀。

热：沉淀时应将溶液加热。

慢：沉淀剂的加入速度要缓慢。

搅：沉淀时要用玻璃棒不断搅拌。

陈：沉淀完全后，要静置一段时间陈化。

为达到上述要求，沉淀操作时，应一手拿滴管，缓慢滴加沉淀剂；另一手持玻璃棒不断搅动溶液，搅拌时玻璃棒不要碰烧杯内壁和烧杯底，速度不宜快，以免溶液溅出。加热应在水浴或电热板上进行，不得使溶液沸腾，否则会引起水溅或产生泡沫飞散，造成被测物损失。沉淀完后，应检查沉淀是否完全，方法是将沉淀溶液静置一段时间，让沉淀下沉。上层溶液澄清后，滴加一滴沉淀剂，观察交界面是否浑浊，如浑浊，表明沉淀未完全，还需加入沉淀剂；如清亮，则表明沉淀完全。

沉淀完全后，盖上表面皿，放置一段时间或在水浴上保温静置 1 h 左右，让沉淀的小晶体生成大晶体，不完整的晶体转为完整的晶体。

2.4.3　过滤和洗涤

过滤和洗涤的目的在于将沉淀从母液中分离出来，使其与过量的沉淀剂及其他杂质组分分开，并通过洗涤将沉淀转化成纯净的单组分。对于需要灼烧的沉淀物，常在玻璃漏斗中用滤纸进行过滤和洗涤；对只需烘干即可称量的沉淀，则在坩埚中进行过滤和洗涤。过滤和洗涤必须一次完成，不能间断。在操作过程中，不得造成沉淀的损失。

1. 滤纸

滤纸分为定性滤纸和定量滤纸两大类。重量分析中使用的是定量滤纸。定量滤纸灼烧后，灰分小于 0.0001 g 的称为“无灰滤纸”，其质量可忽略不计。若灰分质量大于 0.0002 g，则需从沉淀物中扣除其质量。一般市售定量滤纸都已注明每张滤纸的灰分质量，可供参考。定量滤纸一般为圆形，按其孔隙大小分为快速、中速和慢速三种，在过滤时应根据沉淀的性质合理选用。例如，对于晶形沉淀，应选用孔隙小的慢速滤纸；对于无定形沉淀，则应选用孔隙大的快速滤纸。滤纸的大小应根据沉淀量的多少而定，沉淀物的高度不超过滤纸圆锥高度的 1/3 处。

滤纸的折叠：将滤纸对折后再次对折成直角，展开后呈圆锥体，一边一层，另一边三层，放入洁净的漏斗中(滤纸上缘应略低于漏斗上缘)。若滤纸与漏斗不完全密合，可适当调整滤纸的折叠角度至完全密合为止。为使滤纸与漏斗内壁贴合且无气泡，可将三层厚的外层折角撕掉一点并保存在洁净干燥的表面皿上待用。

滤纸的安放：将折叠好的滤纸放入漏斗中，三层处应在漏斗颈出口短的一边，手指按住层厚的一边，用洗瓶吹出少量水将滤纸润湿，然后轻压滤纸除去气泡，使滤纸的锥形上部与漏斗间没有空隙。加水至滤纸边缘，这时漏斗内全部充满水，形成水柱。当漏斗内水全部流尽后，颈内水柱仍能保留且无气泡。若不能形成完整的水柱，可用手指堵住漏斗出口，稍微掀起滤纸三层厚的一边，用洗瓶向滤纸和漏斗间的空隙内注水，直至漏斗颈及锥体的大部分被水充满。然后压紧滤纸边缘，排除气泡，最后缓缓松开堵住漏斗出口的手指，水柱即可形成。在过滤和洗涤过程中，借助水柱的抽吸作用可使过滤速度明显加快。将准备好的漏斗放在漏斗架上，下面放一洁净烧杯盛接滤液，漏斗颈出口长的一边应紧靠杯壁，使滤液沿壁流下以避免冲溅。漏斗位置的高低以过滤时漏斗的出口不接触滤液为宜。

2. 用滤纸过滤

过滤分三步进行：第一步采用倾泻法，尽可能地过滤上层清液；第二步转移沉淀到漏斗上；第三步清洗烧杯和漏斗上的沉淀。此三步操作一定要一次完成，不能间断，尤其是过滤胶状沉淀时更应如此。

倾泻法过滤，即待沉淀沉降后将上层清液沿玻璃棒倾入漏斗内。让沉淀尽可能留在烧杯内，然后再加洗涤液于烧杯中，搅拌沉淀进行充分洗涤，静置澄清后倾出上层清液，这样既可加速过滤，不致堵塞滤纸，又能使沉淀得到充分洗涤。操作时，左手拿盛沉淀的烧杯移至漏斗上方，右手将玻璃棒从烧杯中慢慢取出并在烧杯内壁靠一下，使悬在玻璃棒下端的液滴流入烧杯。慢慢将烧杯倾斜，使上层清液沿玻璃棒缓缓注入漏斗中。倾入的溶液液面至滤纸边缘约 0.5 cm 处时，应暂停倾注，以免沉淀因毛细作用越出滤纸边缘，造成损失。当停止倾注后，将烧杯嘴沿玻璃棒慢慢向上提起，使烧杯直立，再将玻璃棒放回烧杯中以免烧杯嘴上的少量沉淀黏附在玻璃棒上。倾泻法操作如图 2.10 所示。

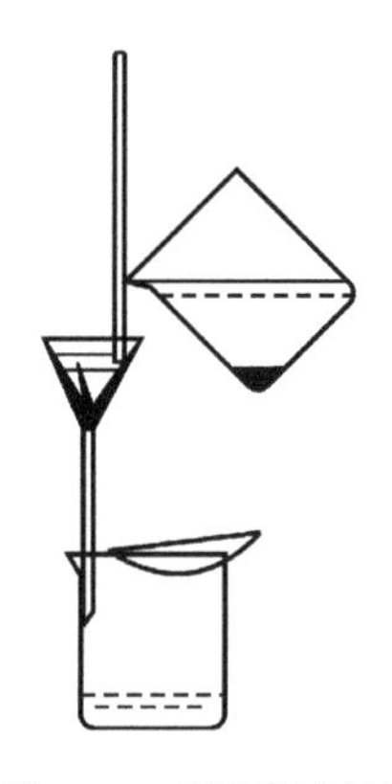

图 2.10　倾泻法过滤

清液倾注完毕后，即可进行初步洗涤。用洗瓶或滴管滴加 15～20 mL 洗涤液或水，从上至下旋转吹洗烧杯内壁及玻璃棒，然后用玻璃棒搅拌沉淀以充分洗涤，再将烧杯倾斜，使沉淀下沉并集中在烧杯一侧，以利于沉淀和清液分离，便于清液的转移。澄清后再倾泻过滤，如此重复过滤，洗涤 3～4 次。

初步洗涤后，即可将沉淀进行定量转移。将少量洗涤液加入盛有沉淀的烧杯中，用玻璃棒充分搅动，并立即将悬浮液转移至滤纸上。对于残留在烧杯内的最后少量沉淀，可反复按以上操作将其完全转移至滤纸上。沉淀完全转移至滤纸上后，在滤纸上进行最后洗涤，用洗瓶吹出细小缓慢的液流，从滤纸上部沿漏斗壁螺旋式向下吹洗，如图 2.11 所示，使沉淀集中到滤纸锥体的底部直到沉淀洗净为止。洗涤数次后，用洁净的表面皿盛接约 1 mL 滤液，选择灵敏、快速的定性反应检验沉淀是否洗净。

图 2.11　沉淀的洗涤

3. 用微孔玻璃漏斗或微孔玻璃坩埚过滤

烘干后即可称量或热稳定性差的沉淀应在玻璃滤器内进行过滤。两种玻璃滤器如图 2.12 所示。这种滤器的滤板是用玻璃粉末在高温下熔结而成的。此类滤器均不能过滤强碱性溶液，以免强碱腐蚀玻璃微孔。按微孔的孔径由大到小可分为六级，即 G1～G6(或称 1 号～6 号)。

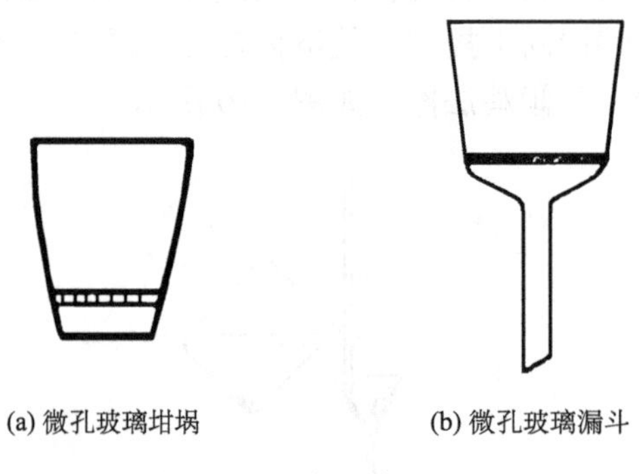

(a) 微孔玻璃坩埚　　(b) 微孔玻璃漏斗

图 2.12　微孔滤器

微孔玻璃滤器必须在抽滤的条件下，采用倾泻法过滤，其过滤、洗涤、转移沉淀等操作均与滤纸过滤法相同。

2.4.4　烘干、灼烧和称量

过滤所得沉淀经加热处理，即获得组成恒定的与化学式表示组成完全一致的沉淀。

1. 沉淀的烘干

烘干一般在 250℃以下进行。凡是用微孔玻璃滤器过滤的沉淀，可用烘干方法处理。将微孔玻璃滤器连同沉淀放在表面皿上，置于烘箱选择合适温度。第一次烘干时间可稍长(如 2 h)，第二次烘干时间可缩短为 40 min。沉淀烘干后，置于干燥器中冷至室温后称量。如此反复操作几次，直至恒量为止。注意每次操作条件要保持一致。

2. 坩埚的准备和干燥器的使用

用钴盐或铁盐溶液在洗净、烘干后的坩埚及盖上写明编号，然后在灼烧沉淀时的温度条件下，预先将空坩埚于高温炉中灼烧 15～30 min 至恒量。将灼烧后的坩埚自然冷却后将其夹入干燥器(干燥器的使用如图 2.13 所示)中。先不要立即盖紧干燥器盖，应等热空气逸出后再盖严。将干燥器冷却至室温后即可称量坩埚，然后灼烧 15～20 min，冷却，称量，直至连续两次称得质量之差不超过 0.2 mg，即可认为坩埚已恒量。

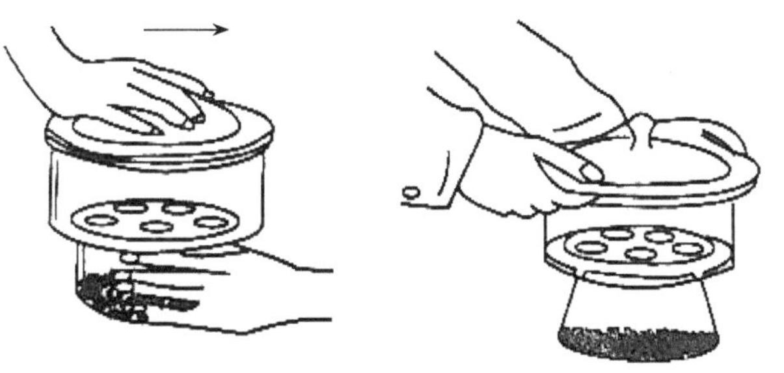

图 2.13　干燥器的使用

3. 沉淀的包裹

用洁净的药铲或顶端扁圆的玻璃棒将滤纸三层部分掀起两层，再用洁净的手指从翘起的滤纸下面将其取出，打开成半圆形，自左端 1/3 半径处向右折叠一次，然后自右端 1/3 半径处向左折叠一次，再自上而下折一次，然后从右向左卷成小卷，如图 2.14(a)所示，最后将其放入已恒量的坩埚内，包裹层数较多的一面朝上，以便于炭化和灰化。对于胶体等蓬松的沉淀，可在漏斗中用玻璃棒将滤纸周边挑起并向内折，把锥体的敞口封住，如图 2.14(b)所示，然后取出倒过来，尖朝上放入坩埚中。

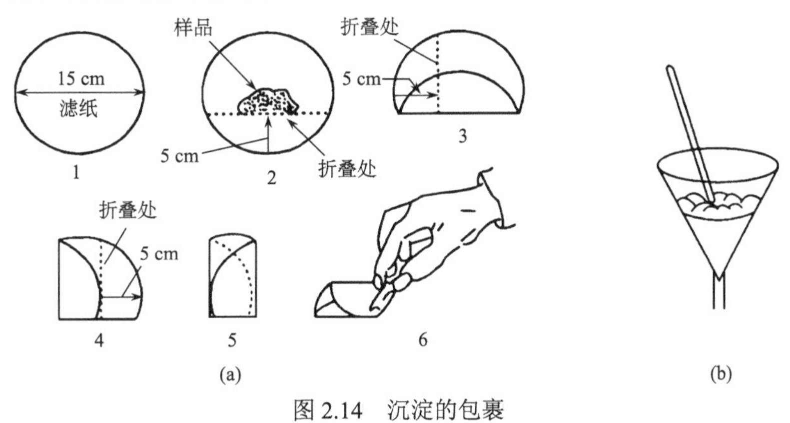

图 2.14　沉淀的包裹

4. 沉淀的烘干、灼烧及称量

将装有沉淀的坩埚置于泥三角上，使多层滤纸部分朝上，将滤纸和沉淀烘干至滤纸全部炭化(滤纸变黑)。炭化后可逐渐升高温度，使滤纸灰化。

沉淀和滤纸灰化后，将坩埚移入高温炉中(根据沉淀性质调节适当温度)，盖上坩埚盖，但留有空隙。在与灼烧空坩埚相同的温度下灼烧 40～45 min，取出，冷至室温，称量。然后进行第二次、第三次灼烧，直至坩埚和沉淀恒量为止。一般第二次以后只需灼烧 20 min 即可。恒量是指相邻两次灼烧后的称量差值不大于 0.4 mg。坩埚每次灼烧完毕从炉内取出后，都应在空气中稍冷后，再移入干燥器中，冷却至室温后称量。然后再灼烧、冷却、称量，直至恒量。

应当注意，每次灼烧、称量和放置的时间都要保持一致。待滤纸全部呈白色后，移至高温炉中灼烧至恒量，然后进行称量。沉淀在坩埚内灼烧的条件及恒量要求应与恒量空坩埚时的相同。

第 3 章　定量分析实验

3.1　基础操作练习

实验 1　分析天平称量练习

实验目的

(1) 熟悉分析天平的使用方法。

(2) 学习常用的称量方法，熟练掌握减量法。

实验原理

分析天平是定量分析实验必备的精密称量仪器。有关分析天平的使用方法和注意事项以及分析试样的称量方法等内容详见 2.1 节。由于分析天平的自重较轻，使用中易因碰撞而发生移动，进而可能造成水平改变，影响称量的准确性，因此操作过程中动作要轻、慢、稳，切不可用力过猛、过快，以免损坏天平。

实验用品

仪器：分析天平(0.1 mg)，烧杯(100 mL)，称量瓶。

试剂：无水 Na_2SO_4(s)。

实验步骤

(1) 固定质量称量法(称取 3 份 0.5000 g 试样)。

打开分析天平，调零后，将洁净、干燥的小烧杯放在秤盘中央(拿取时使用纸带，按规定操作)，关好天平侧门，然后按清零(或 TAR)键。待天平显示 0.0000 g 后，打开天平侧门，用小药匙将试样缓慢加到小烧杯的中央，直到天平显示 0.5000 g。若所称的量小于该值，可继续加试样；若显示的量超过该值，则需重新称量。每次称量后及时将数据记录于表 3.1。

(2) 减量法(称取 3 份 0.45～0.50 g 试样)。

(i) 将盛有试样的称量瓶置于天平秤盘中央，按清零(或 TAR)键，此时显示屏上显示值为 0.0000 g。

(ii) 按正确操作取出称量瓶，在小烧杯的上方缓缓敲出所需质量的试样。待取样完毕后，盖上瓶盖，再将称量瓶放回天平秤盘上，此时的显示值即为取出试样的质量(不考虑负号)。如该质量小于称量范围，可以继续敲出部分试样，直至它的质量在所要求的范围内；如该质量超出称量范围的上限，则不记读数，按清零(或 TAR)键，待显示为 0.0000 g 后，重新进行称量。重复上述操作，每得到一个称量值后都按清零(或 TAR)键，再进行下一次称量，这样就可以连续简便地得到一系列称量值。每次称量后及时在表 3.1 中记录数据。

数据记录

表 3.1 称量练习

	1	2	3
固定质量称量法 m/g			
减量法 m/g			

注意事项

在完成同一个实验内容的过程中，应使用同一台分析天平称量以减小误差。

思考题

(1) 记录称量数据时，应准确至小数点后哪一位？为什么？

(2) 称量时，为什么称量物应放在天平秤盘中央？

(3) 使用称量瓶时，应如何操作才能使试样不致损失？

实验 2 滴定分析基本操作练习

实验目的

(1) 掌握酸式和碱式滴定管的使用方法。

(2) 练习滴定分析的基本操作及常用指示剂的终点判断。

(3) 学习滴定管的正确读数及数据记录。

实验原理

用 HCl 溶液与 NaOH 溶液进行相互滴定的过程中，若采用同一种指示剂，不断改变被滴定溶液的体积，则滴定剂的用量也随之变化，但它们相互反应的体积比 (V_{HCl}/V_{NaOH}) 应基本不变。因此，在 HCl 溶液和 NaOH 溶液准确浓度未知的情况下，通过计算 HCl 溶液和 NaOH 溶液的体积比 (V_{HCl}/V_{NaOH}) 的精密度，可以检查实验者对滴定分析操作和终点判断的掌握情况。

实验用品

仪器：酸式滴定管 (50 mL)，碱式滴定管 (50 mL)，试剂瓶 (500 mL，其中一个带橡皮塞或塑料塞)，锥形瓶 (250 mL)，烧杯 (100 mL)，量筒 (10 mL，100 mL)，分析天平。

试剂：NaOH (AR)，浓盐酸 (AR，1.19 g/mL)，甲基橙水溶液 (0.1%)，酚酞 (0.2%，乙醇溶液)。

除指示剂外，定量分析中所用试剂一般为分析纯，水为蒸馏水或去离子水。

实验步骤

(1) NaOH 溶液 (0.1 mol/L) 的配制。

在分析天平上迅速称取 2.0 g NaOH 固体于 100 mL 烧杯中，加入约 50 mL 水，搅拌使其完全溶解后转移至 500 mL 试剂瓶中，再加 450 mL 左右水，盖紧橡皮塞，摇匀，贴上标签。

(2) HCl 溶液(0.1 mol/L)的配制。

用 10 mL 量筒量取 4.2～4.5 mL 浓盐酸，倒入装有一定体积水的 500 mL 试剂瓶中。用水洗量筒 2～3 次，洗涤液均转入试剂瓶中，最后加水稀释至 500 mL 左右。盖上玻璃塞，摇匀，贴上标签。浓盐酸挥发性很强，以上操作应在通风橱中进行。

(3) 碱式滴定管的操作练习和终点判断。

对滴定管进行检漏并洗涤(包括用蒸馏水洗)完毕后，再用 5～10 mL 待装入的溶液润洗 3 次，装入相应的滴定剂，排尽管下端的气泡，将管内液面调至零刻度线或稍下处，记录初始体积(估读至 0.01 mL)。

从酸式滴定管中放出几毫升 HCl 溶液于 250 mL 锥形瓶中，用 20 mL 左右水稀释，加入 2 滴酚酞指示剂，摇匀。由碱式滴定管中逐滴滴出 NaOH 溶液于锥形瓶中(特别注意练习加一滴和半滴溶液的操作)，观察酚酞指示剂在终点附近变色的情况，滴定至溶液呈微红色且 30 s 不褪色为终点。再用酸式滴定管加入少许 HCl 溶液于上述锥形瓶中，使溶液的红色褪尽，继续用 NaOH 溶液滴定至终点。如此反复练习至能较自如地控制滴定速度并能准确判断终点为止，并进行读数练习(准确读至 0.01 mL)。注意此时滴定管尖端和乳胶管内不得有气泡，并等待一定时间后再读数。

在此基础上，采用累计体积法测定两种溶液用量的体积比(V_{HCl}/V_{NaOH})。步骤如下：从酸式滴定管中放出约 18 mL HCl 溶液于锥形瓶中(每放出一次溶液或滴定后都要准确记录读数)，加入 2 滴酚酞指示剂，用 NaOH 溶液滴定至终点。在前次读数的基础上，由酸式滴定管再放出 2 mL HCl 溶液于同一锥形瓶中，使溶液的微红色褪去，再用 NaOH 溶液继续滴定至终点。整个过程中，两滴定管中均不重新装液，读出累计消耗的滴定剂体积。重复上述步骤，完成第三次滴定。计算酸碱溶液用量的体积比(V_{HCl}/V_{NaOH})。

(4) 酸式滴定管的操作练习和终点判断。

从碱式滴定管中放出几毫升 NaOH 溶液于锥形瓶中，加 20 mL 水稀释，并加入 1～2 滴甲基橙指示剂(练习过程中，随着锥形瓶中溶液体积的增大，可酌情补加指示剂)，摇匀，溶液呈现黄色。用酸式滴定管中的 HCl 溶液进行滴定练习(特别注意练习加一滴和半滴溶液的操作)，观察甲基橙指示剂在终点附近的变色情况，滴定至溶液由黄色恰好变为橙色时为终点。再由碱式滴定管放 1～2 mL NaOH 溶液，使锥形瓶中的溶液重新变为黄色，再用 HCl 溶液滴定。反复练习酸式滴定管的使用方法，学会准确判断终点，并进行读数练习。

在此基础上，采用累计体积法测定两种溶液用量的体积比(V_{HCl}/V_{NaOH})。步骤如下：从碱式滴定管中放出约 18 mL NaOH 溶液于锥形瓶中，加入 2 滴甲基橙指示剂，用 HCl 溶液滴定至终点。在前次读数的基础上，由碱式滴定管再放出 2 mL NaOH 溶液于同一锥形瓶中，使溶液颜色重新变回黄色，再用 HCl 溶液继续滴定至终点。整个过程中，两滴定管中均不重新装液，读出累计消耗的滴定剂体积。重复上述步骤，完成第三次滴定。将消耗的 HCl 溶液和 NaOH 溶液的体积记录于表 3.2 中。

数据记录与处理

计算酸碱溶液用量的体积比(V_{HCl}/V_{NaOH})，要求相对平均偏差 $\bar{d}_r \leqslant 0.3\%$，否则应重新进行滴定。

表 3.2　滴定分析操作练习

	1	2	3
V_{HCl}(终)/mL			
V_{HCl}(始)/mL			
V_{HCl}/mL			
V_{NaOH}(终)/mL			
V_{NaOH}(始)/mL			
V_{NaOH}/mL			
V_{HCl}/V_{NaOH}			
V_{HCl}/V_{NaOH} 平均值			
$\|d_i\|$			
$\bar{d}_r$ /%			

思考题

(1)滴定管使用前为什么要用待装入的溶液充分润洗内壁？所用的锥形瓶是否应烘干后再使用？为什么？

(2)用 NaOH 溶液滴定酸性溶液，以酚酞作指示剂时，为什么要滴定至溶液呈微红色且 30 s 不褪去即为终点？溶液红色褪去的原因是什么？

3.2　酸碱滴定实验

实验 3　NaOH 标准溶液的配制和标定

实验目的

(1)掌握 NaOH 标准溶液的配制和标定方法。

(2)掌握碱式滴定管的使用及酚酞指示剂滴定终点的判断。

实验原理

NaOH 易潮解，且容易与空气中的 CO_2 反应生成 Na_2CO_3，故 NaOH 溶液只能用间接法配制，然后用基准物质标定其准确浓度。

为了配制不含 Na_2CO_3 的 NaOH 标准溶液，常用 NaOH 饱和水溶液。因为 Na_2CO_3 不溶于其中，待 Na_2CO_3 沉淀后，取上层清液，再稀释至所需浓度即可。此外，配制 NaOH 溶液的蒸馏水必须煮沸，以除去其中的 CO_2。

标定碱溶液的基准物质较多，如草酸、苯甲酸、邻苯二甲酸氢钾($KHC_8H_4O_4$，KHP)等。

其中，KHP 具有易制得纯品、不易吸水、稳定、摩尔质量较大等优点。本实验采用 KHP 标定 NaOH 溶液。其反应如下：

$$C_6H_4(COOH)(COOK) + NaOH = C_6H_4(COONa)(COOK) + H_2O$$

由于反应产物为二元弱碱，化学计量点时溶液呈弱碱性(pH ≈ 9)，故应采用酚酞作为指示剂。

计算 NaOH 标准溶液浓度的公式如下：

$$c_{NaOH} = \frac{m_{KHP} \times 1000}{M_{KHP} \times V_{NaOH}} \qquad (M_{KHP} = 204.22\ g/mol)$$

实验用品

仪器：碱式滴定管(50 mL)，锥形瓶(250 mL)，烧杯(250 mL)，移液管(10 mL)，容量瓶(1000 mL)，量筒(50 mL)，分析天平等。

试剂：邻苯二甲酸氢钾(基准物质)，NaOH(AR)，酚酞(0.2%，乙醇溶液)。

实验步骤

(1) NaOH 标准溶液(0.1 mol/L)的配制。

详见实验 2。

(2) NaOH 标准溶液(0.1 mol/L)的标定。

准确称取 0.4～0.5 g 邻苯二甲酸氢钾于 250 mL 锥形瓶中，加 20～30 mL 水，温热使其溶解，冷却后加 2～3 滴酚酞指示剂，用 NaOH 溶液滴定至溶液呈微红色且 30 s 不褪色即为终点。在表 3.3 中记录消耗的 NaOH 溶液的体积。平行滴定 3 次。

数据记录与处理

计算 NaOH 标准溶液的浓度及其相对平均偏差($\bar{d}_r \leqslant 0.3\%$)。

表 3.3　NaOH 溶液的标定

	1	2	3
m_{KHP}/g			
V_{NaOH}(终)/mL			
V_{NaOH}(始)/mL			
V_{NaOH}/mL			
c_{NaOH}/(mol/L)			
$\bar{c}_{NaOH}$/(mol/L)			
$\lvert d_i \rvert$			
$\bar{d}_r$/%			

思考题

(1) 如何计算称取基准物质 KHP 的质量范围？KHP 太多或太少有何影响？

(2) 称取 NaOH 及 KHP 时各选用什么天平？为什么？

(3) 溶解基准物质时加入 20～30 mL 水，是用量筒量取还是用移液管移取？

实验 4　HCl 溶液的配制和标定

实验目的

(1) 掌握 HCl 溶液的配制和标定。

(2) 掌握酸式滴定管的使用及甲基橙指示剂滴定终点的判断。

实验原理

市售浓盐酸为无色透明的 HCl 水溶液，含量为 36%～38%，密度为 1.18 g/L。由于浓盐酸易挥发出 HCl 气体，若直接配制准确度差，因此配制盐酸标准溶液时需用间接配制法。

标定 HCl 标准溶液的浓度，常用的基准物质有无水碳酸钠(Na_2CO_3)和硼砂($Na_2B_4O_7 \cdot 10\,H_2O$)等。本实验采用 Na_2CO_3 为基准物质。

碳酸钠易吸收空气中的水分，先将其于 270～300℃干燥 1 h，然后保存于干燥器中备用。其标定反应为

$$Na_2CO_3 + 2HCl = 2NaCl + H_2O + CO_2\uparrow$$

化学计量点时，溶液为 H_2CO_3 饱和溶液，pH 为 3.9，可以用甲基橙作为指示剂。

计算 HCl 标准溶液浓度的公式如下：

$$c_{HCl} = \frac{2m_{Na_2CO_3} \times 1000}{M_{Na_2CO_3} \times V_{HCl}} \qquad (M_{Na_2CO_3} = 105.99\ g/mol)$$

实验用品

仪器：酸式滴定管(50 mL)，锥形瓶(250 mL)，移液管(25 mL)，试剂瓶，容量瓶(250 mL)，烧杯，量筒(50 mL)，分析天平等。

试剂：浓盐酸(36%～38%，1.18 g/L)，无水碳酸钠(优级纯)，NaOH 溶液(0.1 mol/L)，甲基橙水溶液(0.1%)。

实验步骤

(1) HCl 溶液(0.1 mol/L)的配制。

用量筒量取约 9.0 mL 浓盐酸，倒入装有一定量水的试剂瓶中，加水稀释至 1000 mL，混匀。

(2) HCl 溶液(0.1 mol/L)的标定。

准确称取 0.10～0.12 g 无水碳酸钠置于 250 mL 锥形瓶中，加 20～30 mL 水使其溶解(或准确称取 1.0～1.2 g 无水碳酸钠于烧杯中，加少量水溶解后，转移到 250 mL 容量瓶中，加水

定容，用移液管移取 25.00 mL 于锥形瓶中）。再加 1～2 滴甲基橙指示剂，用 HCl 溶液滴至溶液由黄色变为橙色，即为终点。在表 3.4 中记录消耗的 HCl 溶液的体积。平行滴定 3 次。

数据记录与处理

计算 HCl 标准溶液的浓度及其相对平均偏差（$\bar{d}_r \leqslant 0.3\%$）。

表 3.4　HCl 溶液的标定

	1	2	3
$m_{Na_2CO_3}$/g			
V_{HCl}(终)/mL			
V_{HCl}(始)/mL			
V_{HCl}/mL			
c_{HCl}/(mol/L)			
$\bar{c}_{HCl}$/(mol/L)			
$\lvert d_i \rvert$			
$\bar{d}_r$/%			

注意事项

(1) 无水碳酸钠极易吸水，称量速度要快。

(2) 溶液中 CO_2 过多，酸度增加，会使终点提前。在接近滴定终点时，应剧烈振摇溶液以加快 H_2CO_3 的分解，并加热除去过量的 CO_2，待冷却后再滴定。

思考题

(1) 为什么不能用直接配制法配制 HCl 标准溶液？

(2) 若滴定管未用 HCl 标准溶液进行润洗，对标定结果有何影响？

(3) 用 Na_2CO_3 标定 HCl 溶液是否可以用酚酞作指示剂？

实验 5　食品添加剂中硼酸含量的测定

实验目的

(1) 掌握用酸碱滴定法测定硼酸的原理和方法。

(2) 熟悉碱式滴定管的使用及酚酞指示剂滴定终点的判断。

实验原理

H_3BO_3 的 $K_a = 7.3 \times 10^{-10}$，故不能用 NaOH 标准溶液直接进行滴定。在 H_3BO_3 中加入甘油溶液使其生成甘油硼酸，其 $K_a = 3 \times 10^{-7}$，可用 NaOH 标准溶液直接滴定。反应式为

$$\begin{array}{l}H_2C-OH\\|\\HC-OH\\|\\H_2C-OH\end{array} + H_3BO_3 \rightleftharpoons \begin{array}{l}H_2C-OH\\|\\HC-O\diagdown\\|\qquad\quad BOH\\H_2C-O\diagup\end{array} + 2H_2O$$

$$\begin{array}{l}H_2C-OH\\|\\HC-O\diagdown\\|\qquad\quad BOH\\H_2C-O\diagup\end{array} + NaOH \rightleftharpoons \begin{array}{l}H_2C-OH\\|\\HC-O\diagdown\\|\qquad\quad BONa\\H_2C-O\diagup\end{array} + H_2O$$

化学计量点时，溶液呈弱碱性，可选用酚酞作为指示剂。

实验用品

仪器：碱式滴定管(50 mL)，锥形瓶(250 mL)，烧杯，容量瓶(250 mL)，量筒(25 mL)，分析天平等。

试剂：邻苯二甲酸氢钾(基准物质)，氢氧化钠(AR)，中性甘油(1∶2)，酚酞(0.2%，乙醇溶液)，硼酸样品。

实验步骤

(1) NaOH 溶液(0.1 mol/L)的配制和标定。

详见实验 3。

(2) 硼酸含量的测定。

准确称取 0.29～0.32 g 硼酸样品于 250 mL 锥形瓶中，加 25 mL 中性甘油，温热使其溶解，迅速冷却，加 2～3 滴酚酞指示剂。用 NaOH 溶液(0.1 mol/L)滴定，直到溶液呈微红色且 30 s 不褪色，即为终点。在表 3.5 中记录消耗的 NaOH 溶液的体积。平行滴定 3 次。

数据记录与处理

根据记录的实验数据，分别计算 NaOH 标准溶液的准确浓度和硼酸样品中硼酸的含量、平均值和相对平均偏差($\bar{d}_r \leqslant 0.3\%$)。

表 3.5　硼酸含量的测定

	1	2	3
$m_{H_3BO_3}$ /g			
V_{NaOH}(终)/mL			
V_{NaOH}(始)/mL			
V_{NaOH}/mL			
$w_{H_3BO_3}$ /%			
$\bar{w}_{H_3BO_3}$ /%			
$\lvert d_i \rvert$			
$\bar{d}_r$ /%			

注意事项

(1)加甘油后不要摇动锥形瓶，加热后再摇动。

(2)终点可补加1～2滴酚酞指示剂，增加终点颜色的辨识度。

思考题

(1)为什么不能用直接滴定法测定硼酸的含量？

(2)硼酸的共轭碱是什么？是否可以用酸碱滴定法直接测定硼酸共轭碱的含量？

(3)量取甘油应用什么量器？是否需要用甘油润洗？为什么？

实验6　混合碱的分析

实验目的

(1)掌握双指示剂法测定混合碱的原理和方法。

(2)进一步熟悉酸式滴定管和移液管的操作。

实验原理

工业混合碱是NaOH与Na_2CO_3或者Na_2CO_3与$NaHCO_3$的混合物。采用HCl标准溶液作为滴定剂，先后使用酚酞和甲基橙两种指示剂，在同一份试液中连续滴定，根据消耗的滴定剂的体积，可以判断混合碱的组成，并测定出各组分的含量。这种测定方法称为“双指示剂法”。

在混合碱试液中先加入酚酞指示剂，用HCl标准溶液进行滴定，至试液由紫红色渐变为微红色为第一终点。此时，混合碱中的NaOH已完全反应，而Na_2CO_3只被滴定至$NaHCO_3$，化学计量点溶液pH≈8.32，记录消耗HCl标准溶液的体积为V_1。有关滴定反应为

$$NaOH + HCl = NaCl + H_2O$$

$$Na_2CO_3 + HCl = NaCl + NaHCO_3$$

接着在同一份试液中加入甲基橙指示剂，继续用HCl标准溶液滴定，至试液由黄色突变为橙色时为第二终点。此时，$NaHCO_3$也与HCl反应完毕，化学计量点溶液pH≈3.89。记录消耗HCl标准溶液的体积为V_2。反应如下：

$$NaHCO_3 + HCl = NaCl + H_2O + CO_2\uparrow$$

根据消耗的体积V_1 、V_2，可判断混合碱的组成，并计算各组分的含量。

当$V_1 > V_2$时，试样为NaOH与Na_2CO_3的混合物：

$$\rho_{NaOH} = \frac{c_{HCl} \times (V_1 - V_2) \times 10^{-3} \times M_{NaOH}}{25.00 \times 10^{-3}}$$

$$\rho_{Na_2CO_3} = \frac{c_{HCl} \times V_2 \times 10^{-3} \times M_{Na_2CO_3}}{25.00 \times 10^{-3}}$$

当$V_1 < V_2$时，试样为Na_2CO_3与$NaHCO_3$的混合物：

$$\rho_{Na_2CO_3}=\frac{c_{HCl}\times V_1\times 10^{-3}\times M_{Na_2CO_3}}{25.00\times 10^{-3}}$$

$$\rho_{NaHCO_3}=\frac{c_{HCl}\times (V_2-V_1)\times 10^{-3}\times M_{NaHCO_3}}{25.00\times 10^{-3}}$$

当 $V_1=V_2$ 时，试样为 Na_2CO_3。

如仅需测定工业混合碱的总碱量，则只要加入甲基橙一种指示剂，用 HCl 标准溶液滴定至终点。消耗的总体积应为 V_1+V_2，并将混合碱折算成 Na_2O 的含量计算其总碱量。

实验用品

仪器：酸式滴定管（50 mL），锥形瓶（250 mL），移液管（25 mL），试剂瓶，容量瓶（250 mL），烧杯，量筒，分析天平等。

试剂：HCl 标准溶液（0.1 mol/L），无水 Na_2CO_3（基准物质，270～300℃干燥 1 h，干燥器中保存），酚酞（0.2%，乙醇溶液），甲基橙水溶液（0.1%）。

实验步骤

（1）HCl 溶液（0.1 mol/L）的配制和标定。

详见实验 4。

（2）混合碱的测定（双指示剂法）。

用移液管移取 25.00 mL 混合碱试液于 250 mL 锥形瓶中，加入 2～3 滴酚酞指示剂，用已标定的 HCl 标准溶液滴定，至试液由紫红色变为微红色为第一终点。在表 3.6 中记录消耗 HCl 标准溶液的体积 V_1。

表 3.6　混合碱成分的测定

	1	2	3
V_1(终)/mL			
V_1(始)/mL			
V_1/mL			
V_2(终)/mL			
V_2(始)/mL			
V_2/mL			
ρ_1/(g/L)			
$\bar{\rho}_1$/(g/L)			
$\bar{d}_r$/%			
ρ_2/(g/L)			
$\bar{\rho}_2$/(g/L)			
$\bar{d}_r$/%			
(V_1+V_2)/mL			
$\rho_{总}$/(g/L)			
$\bar{\rho}_{总}$/(g/L)			
$\bar{d}_r$/%			

再在同一份试液中加入 2 滴甲基橙指示剂(试液略显橙色)，继续用上述 HCl 标准溶液滴定，至试液由黄色变为橙色时为第二终点。在表 3.6 中记录第二次消耗 HCl 溶液的体积 V_2。

平行滴定 3 次。

数据记录与处理

(1) 根据 V_1 与 V_2 的相对大小判断混合碱的组成。

(2) 确定各组分消耗 HCl 标准溶液的体积，计算各组分的质量浓度 ρ(g/L)、平均值和相对平均偏差($\bar{d}_r \leqslant 0.3\%$)。

(3) 根据 V_1+V_2(mL) 计算试样的总碱量 Na_2O 的质量浓度(g/L)、平均值和相对平均偏差($\bar{d}_r \leqslant 0.3\%$)。

注意事项

(1) 临近第一终点前，如果滴定速度过快，试液振摇不充分，会造成滴定剂 HCl 局部过浓，致使少量 Na_2CO_3 直接与 HCl 反应并分解成 CO_2 逸出，造成测定误差。

(2) 临近第二终点前，一定要充分振摇试液，避免因 H_2CO_3 过饱和致使试液酸度升高而导致终点提前。

思考题

(1) 用双指示剂法测定混合碱组成的原理和方法是什么？

(2) 判断下列五种情况下混合碱的组成：

①$V_1=0$，$V_2>0$；②$V_1>0$，$V_2=0$；③$V_1>V_2$；④$V_1<V_2$；⑤$V_1=V_2$。

实验 7 铵盐中氮含量的测定

实验目的

(1) 掌握用甲醛法测定铵态氮肥中氮含量的原理和方法。

(2) 熟悉碱式滴定管的操作。

实验原理

NH_4Cl 和 $(NH_4)_2SO_4$ 是常用的氮肥，是强酸弱碱盐。由于铵盐中 NH_4^+ 的酸性太弱($K_a=5.6\times10^{-10}$)，不能用 NaOH 标准溶液直接准确滴定，可采用甲醛法使弱酸强化，反应按下式定量进行：

$$4NH_4^+ + 6HCHO = (CH_2)_6N_4H^+ + 3H^+ + 6H_2O$$

生成的混合酸(其中质子化六次甲基四胺的 $K_a=7.1\times10^{-6}$)可用 NaOH 标准溶液滴定。由于滴定产物中 $(CH_2)_6N_4$ 是弱碱，因此可用酚酞指示终点。甲醛法操作简便快速，用于强酸铵盐中氮含量的测定，但其准确度不如蒸馏法。

实验用品

仪器：碱式滴定管(50 mL)，移液管(25 mL)，锥形瓶(250 mL)，烧杯，分析天平等。

试剂：邻苯二甲酸氢钾(KHP，GR，105～110℃干燥至恒量，干燥器中保存)，NaOH(AR)，甲醛(40%，AR)，酚酞(0.2%，乙醇溶液)，硫酸铵试样。

实验步骤

(1) NaOH 溶液(0.1 mol/L)的配制和标定。

详见实验 3。

(2) 甲醛的中和。

甲醛易被氧化，其中常含有少量甲酸，应事先除去。取原装甲醛(40%)的上层清液于烧杯中，加 2 滴酚酞指示剂，用 NaOH 溶液中和至甲醛溶液呈微红色。

(3) 试样的分析。

准确称取 0.13～0.16 g(可根据 NaOH 溶液的准确浓度进行估算)硫酸铵试样于 250 mL 锥形瓶中，用 20～30 mL 蒸馏水溶解后，加入 5 mL 已中和的甲醛溶液和 2 滴酚酞指示剂，摇匀。静置 1 min 待反应完全后，用已标定的 NaOH 标准溶液滴定试液至微红色且 30 s 不褪色，即为终点。在表 3.7 中记录消耗 NaOH 溶液的体积。平行滴定 3 次。

为了减小称量误差，可准确称取 1.6～1.8 g 硫酸铵试样于烧杯内，加入约 40 mL 水溶解后，定量转入 250 mL 容量瓶中，加蒸馏水稀释、定容并摇匀。用移液管移取 25.00 mL 试液于 250 mL 锥形瓶中。以下操作同上。

数据记录与处理

根据记录的实验数据，分别计算 NaOH 溶液的准确浓度和硫酸铵试样中氮的质量分数、平均值和相对平均偏差($\bar{d}_r \leqslant 0.3\%$)。

表 3.7 铵盐中氮含量的测定

	1	2	3
$m_{试样}$/g			
V_{NaOH}(终)/mL			
V_{NaOH}(始)/mL			
V_{NaOH}/mL			
ρ_N/(g/L)			
$\bar{\rho}_N$/(g/L)			
$\lvert d_i \rvert$			
$\bar{d}_r$/%			

思考题

(1) 能否用甲醛法测定其他铵盐如 NH_4Cl、NH_4NO_3 和 NH_4HCO_3 中的氮含量？为什么？

(2) 本实验称取硫酸铵试样 0.13 g～0.16 g，其称量误差为多少？为了使称量的相对误差不大于 0.1%，应如何操作？

实验 8　磷肥中全磷的测定

实验目的

掌握磷钼酸喹啉容量法测定磷肥中全磷的原理和方法。

实验原理

在酸性条件下，磷肥中的磷转化为磷酸。在含有硝酸的酸性溶液和煮沸的条件下，磷酸与过量的钼酸钠和喹啉生成黄色的磷钼酸喹啉沉淀。将沉淀过滤洗净后，溶于过量的标准碱溶液中，然后用酸标准溶液回滴过量的碱，根据碱和酸的消耗量即可计算五氧化二磷的百分含量。有关反应为

$$H_3PO_4+3C_9H_7N+12Na_2MoO_4+24HNO_3 = (C_9H_7N)_3H_3[P(Mo_3O_{10})_4]\cdot H_2O+11H_2O+24NaNO_3$$

$$(C_9H_7N)_3H_3[P(Mo_3O_{10})_4]\cdot H_2O+26OH^- = HPO_4^{2-}+12MoSO_4^{2-}+3C_9H_7N+15H_2O$$

过量：$OH^-+H^+ = H_2O$

实验用品

仪器：碱式滴定管(50 mL)，电炉，容量瓶(250 mL)，移液管(25 mL)，量筒(50 mL)，洗耳球，滤纸，烧杯，抽滤装置，分析天平。

试剂：NaOH 溶液(4 g/L)，NaOH 标准溶液(0.5 mol/L)，HCl 标准溶液(0.25 mol/L)，溴百里酚蓝-酚酞指示剂，高喹试剂，盐酸(1 : 1)，硝酸(1 : 1)，磷肥试样。

实验步骤

(1)样品分解。

称取 1 g 试样于 250 mL 烧杯中，用少量水(约 5 mL)润湿，加入 20～25 mL 盐酸和 7～9 mL 硝酸，混匀，盖上表面皿。在电炉上缓慢加热 30 min(在加热过程中可补充水，防止烧干)。取下烧杯，待冷却后加入 10 mL 盐酸(1 : 1)，减压过滤，用水洗涤烧杯及滤纸 5 次，每次约 10 mL，将滤液转移至 250 mL 容量瓶中，冷却，稀释至刻度。

(2)沉淀的形成。

取 25.00 mL 试液于 300 mL 烧杯中，加入 10 mL 硝酸(1 : 1)溶液，用水稀释至 100 mL，盖上表面皿，预热近沸。加入 50 mL 高喹试剂，微沸搅拌 1 min(此时生成磷钼酸喹啉沉淀)，冷却至室温。

(3)沉淀的过滤及洗涤。

将上述得到的混合物进行减压过滤，先将上层清液滤完，然后用倾泻法洗涤沉淀 3～4 次，每次用蒸馏水约 25 mL，再将沉淀移入滤器中，用蒸馏水洗 3～5 次。

(4)沉淀的溶解及滴定。

将沉淀连同滤纸移入原烧杯中，用滴定管滴加 NaOH 标准溶液，将粘在布氏漏斗中的黄色固体溶解洗入烧杯中，待黄色消失后用水冲洗漏斗，收集洗涤液加入烧杯。将滤纸捣碎，继续滴加 NaOH 标准溶液并充分搅拌至黄色沉淀完全溶解。加入 100 mL 新煮沸的水，搅匀

溶液后加入 1 mL 溴百里酚蓝-酚酞指示剂，用 HCl 标准溶液滴定至溶液从紫色经灰蓝变为淡黄色(或无色)为终点。

(5)空白实验。

按照上述步骤，平行做空白实验。

数据记录与处理

根据记录的实验数据，计算五氧化二磷的百分含量。

$$w_{P_2O_5}=\frac{1}{52}\times\frac{c_{NaOH}(V_1-V_3)-c_{HCl}(V_2-V_4)}{m\dfrac{25.00}{250.0}\times 1000}\times M_{P_2O_5}$$

式中，c_{NaOH} 为 NaOH 标准溶液的浓度，mol/L；c_{HCl} 为 HCl 标准溶液的浓度，mol/L；V_1 为消耗 NaOH 标准溶液的体积，mL；V_2 为消耗 HCl 标准溶液的体积，mL；V_3 为空白实验消耗 NaOH 标准溶液的体积，mL；V_4 为空白实验消耗 HCl 标准溶液的体积，mL；m 为磷肥试样的质量，g；25.00 为吸取试样溶液的体积，mL；250.0 为试样溶液的总体积，mL；$M_{P_2O_5}$ 为五氧化二磷的摩尔质量，141.9 g/mol；1/52 为五氧化二磷与氢氧化钠的摩尔比。

注意事项

磷钼杂多酸只有在酸性条件下才稳定，在碱性溶液中重新分解为原来的简单酸根离子。酸度、温度、配位酸根的浓度都严重影响杂多酸的组成。因此，沉淀条件必须严格控制。理论上，酸度大对沉淀有利。但酸度过高，沉淀困难。酸度过低，反应不完全，造成测量误差。

思考题

(1)解释磷钼酸喹啉容量法测定磷肥中磷的原理。

(2)如何控制磷钼酸喹啉容量法测定磷肥中磷的测定条件？

实验 9　食用醋总酸度的测定

实验目的

(1)了解强碱滴定弱酸过程中溶液 pH 的变化以及指示剂的选择。

(2)学习食用醋中总酸度的测定方法。

实验原理

食用醋的主要酸性物质是乙酸(HAc)，此外还含有少量其他弱酸，如乳酸、琥珀酸、苹果酸、氨基酸等。乙酸的解离常数 $K_a=1.8\times10^{-5}$，可用 NaOH 标准溶液滴定。化学计量点时溶液 pH≈8.7，可选用酚酞为指示剂，滴定终点时溶液由无色变为微红色。滴定时，HAc 和食用醋中可能存在的其他酸均与 NaOH 反应，故滴定所得为总酸度，以 ρ_{HAc}(g/L)表示。

实验用品

仪器：碱式滴定管(50 mL)，移液管(25 mL)，容量瓶(250 mL)，锥形瓶(250 mL)。

试剂：NaOH 标准溶液(0.1 mol/L)，邻苯二甲酸氢钾(KHP，基准物质)，酚酞指示剂(2 g/L，乙醇溶液)，食用醋。

实验步骤

(1) NaOH 溶液(0.1 mol/L)的标定。

详见实验 3。

(2) 食用醋总酸度的测定。

准确移取 25.00 mL 食用醋于 250 mL 容量瓶中，用新煮沸并冷却的蒸馏水稀释至刻度，摇匀。用移液管移取 25.00 mL 上述稀释后的试液于 250 mL 锥形瓶中，加入 2～3 滴酚酞指示剂。用 NaOH 标准溶液(0.1 mol/L)滴至溶液呈微红色且 30 s 内不褪色，即为终点。将消耗 NaOH 标准溶液的体积记录在表 3.8 中。平行滴定 3 次。

数据记录与处理

(1) 写出有关计算公式，计算 NaOH 标准溶液的浓度和相对平均偏差。

(2) 根据 NaOH 标准溶液的浓度和消耗的体积，计算食用醋总酸度、平均偏差和相对平均偏差($\bar{d}_r \leqslant 0.3\%$)。

表 3.8 食用醋总酸度的测定

	1	2	3
$V_{醋}$/mL			
$V_{稀释后}$/mL			
V_{NaOH}(终)/mL			
V_{NaOH}(始)/mL			
V_{NaOH}/mL			
ρ_{HAc}/(g/L)			
$\bar{\rho}_{HAc}$/(g/L)			
$\lvert d_i \rvert$			
$\bar{d}_r$/%			

思考题

测定乙酸含量时，所用蒸馏水不能含二氧化碳，为什么？

实验 10 有机酸摩尔质量的测定

实验目的

(1) 掌握有机酸摩尔质量的测定方法。

(2) 熟悉 NaOH 标准溶液的配制和标定方法。

实验原理

大多数有机酸是弱酸，如草酸（pK_{a1}=1.23，pK_{a2}=4.19）、酒石酸（pK_{a1}=2.85，pK_{a2}=4.34）、柠檬酸（pK_{a1}=3.15，pK_{a2}=4.77，pK_{a3}=6.39）等，它们在水中有一定溶解性。若酸浓度达 0.1 mol/L 左右，且 $cK_a \geqslant 10^{-8}$，则可用 NaOH 标准溶液滴定，常选用酚酞作指示剂，滴定至终点溶液呈微红色，根据 NaOH 标准溶液的浓度和滴定时消耗的体积及称取的纯有机酸的质量，可计算该有机酸的摩尔质量。当有机酸为多元酸时，应根据每一级酸能否被准确滴定的判别式（$c_{ai}K_{ai} \geqslant 10^{-8}$），以及相邻两级酸之间能否分级滴定的判别式（$c_{ai}K_{ai}/c_{ai+1}K_{ai+1} \geqslant 10^5$）判断多元酸与 NaOH 之间反应的计量关系，据此计算出有机酸的摩尔质量。

实验用品

仪器：碱式滴定管（50 mL），移液管（25 mL），容量瓶（250 mL），锥形瓶（250 mL），烧杯，分析天平。

试剂：NaOH 标准溶液（0.1 mol/L），邻苯二甲酸氢钾（KHP，基准物质），酚酞指示剂（2 g/L，乙醇溶液），有机酸（如草酸、酒石酸、柠檬酸、乙酰水杨酸等）。

实验步骤

（1）NaOH 溶液（0.1 mol/L）的配制和标定。

详见实验 3。

（2）有机酸摩尔质量的测定。

准确称取有机酸试样于干燥烧杯中，加蒸馏水溶解，定量转入 250 mL 容量瓶中，用蒸馏水稀释至刻度，摇匀。移取 25.00 mL 溶液于 250 mL 锥形瓶中，加 2～3 滴酚酞指示剂，用 NaOH 标准溶液滴至溶液刚好由无色变为粉红色且 30 s 内不褪色，即为终点。平行滴定 3 次。将消耗 NaOH 标准溶液的体积记录在表 3.9 中。

数据记录与处理

（1）写出有关计算公式，计算 NaOH 标准溶液的浓度和相对平均偏差。

（2）根据 NaOH 标准溶液的浓度和消耗的体积，计算有机酸的摩尔质量、平均偏差和相对平均偏差。

表 3.9　有机酸摩尔质量的测定

	1	2	3
$m_{有机酸}$/g			
$V_{试液}$/mL			
V_{NaOH}(终)/mL			
V_{NaOH}(始)/mL			
V_{NaOH}/mL			
$M_{有机酸}$/(g/mol)			
$\bar{M}_{有机酸}$/(g/mol)			
$\lvert d_i \rvert$			
$\bar{d}_r$/%			

思考题

(1) 如果 NaOH 标准溶液在保存过程中吸收了空气中的二氧化碳，用此标准溶液滴定同一浓度的 HCl 溶液时，分别选用甲基橙和酚酞为指示剂有何区别？为什么？

(2) 草酸、柠檬酸、酒石酸等有机多元酸能否用 NaOH 标准溶液滴定？

3.3　络合滴定实验

实验 11　EDTA 标准溶液的配制、标定和水硬度的测定

实验目的

(1) 掌握 EDTA 标准溶液的配制和标定方法。

(2) 掌握铬黑 T 指示剂和钙指示剂指示终点的判断方法。

(3) 掌握用络合滴定法测定水硬度的原理和方法。

实验原理

乙二胺四乙酸(简称 EDTA，简写为 H_4Y)分子中的四个羧基氧和两个氨基氮可以与金属配位成键，是分析化学中使用最广泛的螯合剂。EDTA 难溶于水，故 EDTA 标准溶液通常用乙二胺四乙酸二钠盐($Na_2H_2Y \cdot 2H_2O$，一般也简称为 EDTA，或 EDTA 二钠盐)配制。乙二胺四乙酸二钠盐很难得到纯物质，其标准溶液用间接法配制。

水的总硬度是指水中 Ca^{2+}、Mg^{2+}的总量，分为钙硬(由 Ca^{2+}引起的硬度)和镁硬(由 Mg^{2+}引起的硬度)。在工业用水中，水的硬度是形成锅垢和影响产品质量的主要因素。因此，水硬度的测定可以为确定用水质量和水处理提供依据。水的硬度常以水中 Ca、Mg 总量换算为 CaO 含量的方法表示，单位为 mg/L。

测定水硬度时，常用 $CaCO_3$ 基准物质标定 EDTA 溶液的浓度，选用钙指示剂(In)指示终点，用 NaOH 溶液控制 pH 为 12～13，溶液由紫红色变为蓝色，即为终点。其变色原理为

滴定前　$Ca + In(蓝色) = CaIn(紫红色)$

滴定中　$Ca + Y = CaY$

终点时　$CaIn(紫红色) + Y = CaY + In(蓝色)$

水的总硬度一般在 NH_3-NH_4Cl 缓冲溶液($pH \approx 10$)中测定，以铬黑 T 作指示剂，用 EDTA 标准溶液直接滴定水中 Ca^{2+}、Mg^{2+}的总量。水样中 Fe^{3+}、Al^{3+}、Cu^{2+}、Pb^{2+}和 Zn^{2+}等干扰离子可以用三乙醇胺掩蔽。

如需分别测定水的钙硬度和镁硬度，可加 NaOH 溶液调节水样的 pH 为 12～13，使 Mg^{2+}形成 $Mg(OH)_2$ 沉淀，以钙指示剂指示终点(紫红色→蓝色)，用 EDTA 标准溶液滴定水样中的钙含量。镁含量即可由钙镁总量与钙含量之差求得。

实验用品

仪器：碱式滴定管(50 mL)，烧杯(100 mL，500 mL)，容量瓶(250 mL)，锥形瓶(250 mL)，

移液管(25 mL)，分析天平。

试剂：乙二胺四乙酸二钠($Na_2H_2Y \cdot 2H_2O$，AR)，$CaCO_3$(GR)，铬黑 T(EBT)指示剂，钙指示剂，NH_3-NH_4Cl 缓冲溶液(pH ≈ 10)，HCl 溶液(6 mol/L)，NaOH 溶液(1 mol/L)，三乙醇胺溶液(1∶2)。

实验步骤

(1) EDTA 标准溶液(0.02 mol/L)的配制。

称取 4.0 g $Na_2H_2Y \cdot 2H_2O$ 于 500 mL 烧杯中，加 200 mL 水微热使其完全溶解。冷却后，转入试剂瓶(如需保存，则用聚乙烯瓶)中，稀释至 500 mL，摇匀，贴上标签。

(2) 钙标准溶液(0.02 mol/L)的配制。

准确称取 0.50～0.55 g 基准物质 $CaCO_3$，置于 100 mL 烧杯中，加几滴水润湿，盖上表面皿，慢慢滴加 5 mL HCl 溶液(6 mol/L)使其溶解，定量转移至 250 mL 容量瓶中，加蒸馏水稀释至刻度，摇匀。

(3) EDTA 标准溶液的标定。

准确移取 25.00 mL 钙标准溶液于 250 mL 锥形瓶中，加入 5 mL NaOH 溶液(1 mol/L)和适量钙指示剂，摇匀，用 EDTA 标准溶液进行滴定，至溶液由紫红色恰好变为蓝色为终点。记录消耗 EDTA 标准溶液的体积(V_0)于表 3.10 中，平行滴定 3 次。

(4) 总硬度的测定。

移取 25.00 mL 水样于 250 mL 锥形瓶中，加入 5 mL 氨性缓冲溶液和 3～4 滴 EBT 指示剂，摇匀，用 EDTA 标准溶液缓慢滴定至溶液由紫红色变为蓝色。记录消耗 EDTA 标准溶液的体积(V_1)于表 3.11 中，平行滴定 3 次。

(5) 钙硬度的测定。

用移液管移取 25.00 mL 水样于 250 mL 锥形瓶中，加入 5 mL NaOH 溶液(1 mol/L)和适量钙指示剂，摇匀，用 EDTA 标准溶液缓慢滴定至溶液由紫红色变为蓝色。记录消耗 EDTA 标准溶液的体积(V_2)于表 3.12 中，平行滴定 3 次。

数据记录与处理

(1) 计算 EDTA 标准溶液的浓度、平均值和相对平均偏差($\bar{d}_r \leqslant 0.3\%$)。

(2) 分别计算水样的总硬度(mg/L)、钙硬度和镁硬度(mg/L)，平均值和相对平均偏差($\bar{d}_r \leqslant 0.3\%$)。

$$\rho_{总} = \frac{c_{EDTA} \times V_1 \times M_{CaO}}{V_{水}} \times 1000$$

$$\rho_{Ca} = \frac{c_{EDTA} \times V_2 \times M_{Ca}}{V_{水}} \times 1000$$

$$\rho_{Mg} = \frac{c_{EDTA} \times (V_1 - V_2) \times M_{Mg}}{V_{水}} \times 1000$$

表 3.10　EDTA 标准溶液的标定

	1	2	3
m_{CaCO_3} /g			
$c_{Ca^{2+}}$ /(mol/L)			
V_0(终)/mL			
V_0(始)/mL			
V_0/mL			
c_{EDTA}/(mol/L)			
$\bar{c}_{EDTA}$/(mol/L)			
$\lvert d_i \rvert$			
$\bar{d}_r$ /%			

表 3.11　总硬度的测定

	1	2	3
$V_{水样}$/mL			
V_1(终)/mL			
V_1(始)/mL			
V_1/mL			
$\rho_{总}$ /(mg/L)			
$\bar{\rho}_{总}$ /(mg/L)			
$\lvert d_i \rvert$			
$\bar{d}_r$ /%			

表 3.12　钙硬度和镁硬度的测定

	1	2	3
$V_{水样}$/mL			
V_2(终)/mL			
V_2(始)/mL			
V_2/mL			
ρ_{Ca}/(mg/L)			
$\bar{\rho}_{Ca}$/(mg/L)			
$\lvert d_i \rvert$			
$\bar{d}_r$ /%			
ρ_{Mg}/(mg/L)			
$\bar{\rho}_{Mg}$/(mg/L)			
$\lvert d_i \rvert$			
$\bar{d}_r$ /%			

思考题

(1)络合滴定中为什么要加入缓冲溶液?

(2) 使用 EBT 指示剂时，溶液 pH 应控制为多少？如何控制？

(3) 本实验滴定时要缓慢进行，为什么？

实验 12　工业原料中钙、镁含量的测定

实验目的

(1) 掌握络合滴定法测定石灰石中钙、镁含量的原理和方法。

(2) 了解络合滴定中指示剂的选择和应用。

实验原理

石灰石是重要的工业原料，其主要成分为 $CaCO_3$，还含有一定量的 $MgCO_3$ 及少量的 Al、Fe、Si 等杂质。其中，钙、镁含量可采用络合滴定法进行测定。

试样经酸溶解后，Ca^{2+}、Mg^{2+}共存于溶液中，Fe^{3+}、Al^{3+}等干扰离子可用三乙醇胺掩蔽。调节溶液 pH ≥12，使 Mg^{2+}生成 $Mg(OH)_2$ 沉淀。用 EDTA 标准溶液滴定，加入钙指示剂指示终点，溶液由紫红色变为蓝色。根据消耗的 EDTA 体积可计算钙的含量(以 CaO 的质量分数表示)。反应式如下：

滴定前　　$Ca + In(蓝色) = CaIn(紫红色)$

滴定中　　$Ca + Y = CaY$

终点时　　$CaIn(紫红色) + Y = CaY + In(蓝色)$

另取一份试液，用三乙醇胺将 Fe^{3+}、Al^{3+}等干扰离子掩蔽后，调节溶液 pH ≈ 10，以铬黑 T(EBT) 为指示剂，用 EDTA 标准溶液滴定至溶液由红色(经紫蓝色)转变为蓝色，即为终点。根据消耗的 EDTA 标准溶液的体积，可计算试样中钙、镁的总量。镁含量可用钙镁总量与钙含量之差求得，以 MgO 的质量分数表示。反应原理如下：

滴定前　　$Mg + EBT(蓝色) = Mg\text{-}EBT(紫红色)$

滴定中　　$Ca(Mg) + Y = CaY(MgY)$

终点时　　$Mg\text{-}EBT(紫红色) + Y = MgY + EBT(蓝色)$

实验用品

仪器：酸式滴定管(50 mL)，锥形瓶(250 mL)，移液管(25 mL)，容量瓶(250 mL)，量筒(10 mL，25 mL)，烧杯(100 mL)，分析天平。

试剂：EDTA 标准溶液(0.02 mol/L)，NaOH 溶液(10%)，HCl 溶液(1∶1)，三乙醇胺溶液(1∶2)，$NH_3\text{-}NH_4Cl$ 缓冲溶液(pH ≈ 10)，钙指示剂，铬黑 T 指示剂，石灰石试样。

实验步骤

(1) EDTA 标准溶液(0.02 mol/L)的配制和标定。

详见实验 11。

(2) 试液的制备。

准确称取 0.25～0.30 g 试样于 100 mL 烧杯中，加少量水润湿，盖上表面皿，慢慢滴加 4～

6 mL HCl 溶液(1∶1)。反应剧烈时稍停，略转动烧杯，使试样完全溶解。用少量水清洗表面皿凸面和烧杯内壁，洗涤液全部倒入烧杯中。将试样溶液全部转移到 250 mL 容量瓶中，加水稀释、定容、摇匀。

(3) 钙含量的测定。

移取 25.00 mL 试液于 250 mL 锥形瓶中，加 20 mL 水和 5 mL 三乙醇胺溶液，摇匀。再加入 10 mL NaOH 溶液(10%)，使溶液 pH 为 12～14。加入少量钙指示剂，摇匀，用 EDTA 标准溶液进行滴定，并不断振荡。因反应速度较慢，临近终点时应慢滴多摇，至溶液由紫红色变为蓝色，即为终点。在表 3.13 中记录消耗 EDTA 标准溶液的体积(V_1)，平行滴定 3 次。

(4) 钙镁总量的测定。

移取 25.00 mL 试液于 250 mL 锥形瓶中，加 20 mL 水和 5 mL 三乙醇胺溶液，再加约 10 mL NH_3-NH_4Cl 缓冲溶液和 3～5 滴 EBT 指示剂，摇匀。用 EDTA 标准溶液滴定，至溶液由紫红色变为蓝色，即为终点。在表 3.13 中记录消耗 EDTA 标准溶液的体积(V_2)，平行滴定 3 次。

数据记录与处理

计算石灰石试样中 CaO、MgO 的质量分数、平均值和相对平均偏差($\bar{d}_r \leqslant 0.3\%$)。

$$w_{CaO}=\frac{c_{EDTA}\times V_1\times M_{CaO}}{25.00\times m_{试样}}\times 250\times 10^3\times 100\%$$

$$w_{MgO}=\frac{c_{EDTA}\times (V_2-V_1)\times M_{MgO}}{25.00\times m_{试样}}\times 250\times 10^3\times 100\%$$

表 3.13 钙、镁含量的测定

	1	2	3
$m_{试样}$/g			
V_1(终)/mL			
V_1(始)/mL			
V_1/mL			
w_{CaO}/%			
$\bar{w}_{CaO}$/%			
$\lvert d_i \rvert$			
$\bar{d}_r$/%			
V_2(终)/mL			
V_2(始)/mL			
V_2/mL			
w_{MgO}/%			
$\bar{w}_{MgO}$/%			
$\lvert d_i \rvert$			
$\bar{d}_r$/%			

思考题

(1) 试说明络合滴定法测定石灰石中钙、镁含量的原理。

(2) 测定钙、镁含量时为什么要加入三乙醇胺？可否在加入缓冲溶液后再加入三乙醇胺？为什么？

实验 13　铅、铋混合溶液中铅、铋含量的测定

实验目的

(1) 理解通过控制溶液酸度提高络合滴定选择性的原理。

(2) 掌握用 EDTA 连续滴定多种金属离子混合试液的方法。

(3) 熟悉二甲酚橙(XO)指示剂的应用和终点的判断。

实验原理

Bi^{3+}和 Pb^{2+}均能与 EDTA 形成稳定的络合物，但它们的形成常数差别较大($\lg K_{BiY} = 27.94$，$\lg K_{PbY} = 18.04$)，符合混合离子分步滴定的条件(当 $c_M = c_N$，$\Delta pM = \pm 0.2$，欲 $|E_t| \leqslant 0.1\%$，则需要$\Delta\lg K \geqslant 6$)。因此，可用二甲酚橙(XO)为指示剂，通过控制混合溶液的酸度，在同一份试液中先后对 Bi^{3+}、Pb^{2+}进行连续滴定。

先调节溶液 $pH \approx 1$，Bi^{3+}与 XO 形成紫红色络合物(Pb^{2+}在此条件下不干扰)，用 EDTA 标准溶液滴定 Bi^{3+}，溶液由紫红色经红色、橙色变为亮黄色，即为滴定的第一终点。然后加入六次甲基四胺调节溶液 pH 为 5～6，此时 Pb^{2+}与 XO 形成紫红色络合物，继续用 EDTA 标准溶液滴定至溶液由紫红色变为亮黄色，即为滴定的第二终点。

为了使标定和测定在相同的反应条件下进行，用基准物质 $ZnSO_4 \cdot 7H_2O$ 标定 EDTA 溶液的浓度，二甲酚橙为指示剂。滴定在 pH 为 5～6 的 $HCl\text{-}(CH_2)_6N_4$ 缓冲溶液中进行，终点时溶液颜色的变化同上。

实验用品

仪器：酸式滴定管(50 mL)，锥形瓶(250 mL)，移液管(25 mL)，容量瓶(250 mL)，烧杯(100 mL)，分析天平等。

试剂：EDTA 标准溶液(0.02 mol/L)，$ZnSO_4 \cdot 7H_2O$ (基准物质)，HNO_3 溶液(0.1 mol/L)，HCl 溶液(1 : 5)，六次甲基四胺溶液(20%，AR)，二甲酚橙溶液(0.2%)，Bi^{3+}、Pb^{2+} 混合溶液($c_{Bi^{3+}} \approx 0.01$ mol/L，$c_{Pb^{2+}} \approx 0.01$ mol/L，$c_{HNO_3} \approx 0.15$ mol/L)。

实验步骤

(1) Zn 标准溶液(0.02 mol/L)的配制。

准确称取 1.40～1.45 g $ZnSO_4 \cdot 7H_2O$ 基准物质于 100 mL 烧杯中，加水溶解，定量转移入 250 mL 容量瓶中，加水稀释、定容、摇匀。

(2) EDTA 标准溶液(0.02 mol/L)的配制和标定。

EDTA 标准溶液的配制详见实验 11。

准确移取25.00 mL Zn标准溶液于250 mL锥形瓶中，加入2 mL HCl溶液(1∶5)和2滴二甲酚橙指示剂，滴加六次甲基四胺溶液至溶液呈稳定的紫红色后，再过量5 mL，摇匀。用待标定的EDTA标准溶液滴定至溶液由紫红色至亮黄色为终点(临近终点时，慢滴多摇，以免过量)。在表3.14中记录V_{EDTA}。平行滴定3次。

(3)铅、铋含量的连续测定。

移取25.00 mL Bi^{3+}、Pb^{2+}混合溶液于250 mL锥形瓶中，加入10 mL HNO_3溶液(0.1 mol/L)和2滴二甲酚橙指示剂，摇匀。用EDTA标准溶液滴定至溶液由紫红色变为亮黄色，即为第一终点。记录消耗的EDTA标准溶液的体积(V_{Bi})。由于Bi^{3+}与EDTA反应的速度较慢，故临近终点时速度不宜过快，且应用力振荡溶液。

酌情向溶液中补加1滴二甲酚橙指示剂，然后滴加六次甲基四胺溶液至溶液呈稳定的紫红色后，再过量5 mL，此时溶液pH应为5～6。继续用EDTA标准溶液滴定至溶液由紫红色变为亮黄色，即为第二终点。在表3.15中记录消耗的EDTA标准溶液的体积$V_{总}$($V_{Pb}=V_{总}-V_{Bi}$)。平行滴定3次。

表3.14 EDTA标准溶液的标定

	1	2	3
$m_{ZnSO_4\cdot 7H_2O}$/g			
V_{EDTA}(终)/mL			
V_{EDTA}(始)/mL			
V_{EDTA}/mL			
c_{EDTA}/(mol/L)			
$\bar{c}_{EDTA}$/(mol/L)			
$\|d_i\|$			
$\bar{d}_r$/%			

表3.15 铅、铋含量的测定

	1	2	3
V_{Bi}(终)/mL			
V_{Bi}(始)/mL			
V_{Bi}/mL			
$V_{总}$/mL			
V_{Pb}/mL			
ρ_{Bi}/(g/L)			
$\bar{\rho}_{Bi}$/(g/L)			
$\|d_i\|$			
$\bar{d}_r$/%			
ρ_{Pb}/(g/L)			
$\bar{\rho}_{Pb}$/(g/L)			
$\|d_i\|$			
$\bar{d}_r$/%			

数据记录与处理

(1) 计算 EDTA 标准溶液的浓度、平均值和相对平均偏差($\bar{d}_r \leqslant 0.2\%$)。

(2) 计算溶液中铅、铋的含量(g/L)、平均值和相对平均偏差($\bar{d}_r \leqslant 0.2\%$)。

注意事项

滴定 Bi^{3+}时，若酸度过低，Bi^{3+} 将水解产生白色浑浊物，使终点过早出现，产生回红现象。此时应放置片刻，继续滴定至溶液变为稳定的透明亮黄色，即为终点。

思考题

(1) 滴定 Bi^{3+}，Pb^{2+} 时，溶液 pH 控制在什么范围？如何调节 pH？

(2) 滴定 Pb^{2+} 时要调节溶液 pH 为 5～6，为什么加入六次甲基四胺而不是乙酸钠？

(3) 滴定 Bi^{3+}之前，加入 HNO_3 的作用是什么？

实验 14　胃舒平片中铝和镁含量的测定

实验目的

(1) 了解成品药剂中组分含量测定的预处理方法。

(2) 掌握返滴定法的原理和操作。

实验原理

胃舒平是一种常见胃药，主要成分为氢氧化铝[$Al(OH)_3$]、三硅酸镁($2MgO\cdot 3SiO_2\cdot xH_2O$)及少量颠茄浸膏和糊精。药片中铝和镁的含量可以用络合滴定法测定。先将药片用硝酸溶解，分离出不溶物质，再加入过量且一定量的 EDTA 溶液，调节 pH 为 3～4，煮沸数分钟使 Al^{3+} 与 EDTA 充分反应。冷却后再将其 pH 调到 5～6，以二甲酚橙为指示剂，用 Zn^{2+}标准溶液返滴定过量的 EDTA，求得氢氧化铝的含量。

另取试液，调节 pH 为 8～9，将 Al^{3+}以沉淀形式分离。在 pH 约为 10 的条件下，以铬黑 T 为指示剂，用 EDTA 滴定滤液中的 Mg^{2+}。

实验用品

仪器：酸式滴定管(50 mL)，烧杯(100 mL，500 mL)，容量瓶(250 mL)，锥形瓶(250 mL)，移液管(10 mL，25 mL)，量筒(10 mL，50 mL)，分析天平，电炉，研钵等。

试剂：EDTA 标准溶液(0.02 mol/L)，锌标准溶液(0.02 mol/L，详见实验 13)，六次甲基四胺(20%)，NH_3H_2O-NH_4Cl 缓冲溶液(pH ≈ 10)，三乙醇胺溶液(1∶2)，HCl 溶液(1∶1)，NaOH 溶液(1 mol/L)，氨水(1∶1)，甲基红指示剂(0.2%，乙醇溶液)，EBT 指示剂，二甲酚橙指示剂(0.2%)。

实验步骤

(1) EDTA 标准溶液(0.02 mol/L)的配制和标定。

详见实验 13。

(2) 样品处理。

取 10 片胃舒平药片研细，称取 2 g 左右粉末于 250 mL 烧杯中，加入 20 mL HCl 溶液(1 : 1)，加蒸馏水稀释至 100 mL，煮沸。冷却后过滤，并用蒸馏水洗涤沉淀。收集滤液及洗涤液于 250 mL 容量瓶中，用蒸馏水稀释至刻度，摇匀。

(3) 铝含量的测定。

准确吸取 10.00 mL 上述试液于 250 mL 锥形瓶中，加蒸馏水稀释至 25 mL 左右，滴加 $NH_3 \cdot H_2O$ 溶液(1 : 1)至刚出现浑浊，再加 HCl 溶液(1 : 1)至沉淀恰好溶解。加入 25.00 mL EDTA 标准溶液，煮沸 1 min 并冷却后，再加入 10 mL 六次甲基四胺溶液(20%)，加 2～3 滴二甲酚橙指示剂，用锌标准溶液滴定至溶液由黄色变为红色为终点。平行滴定 3 次。在表 3.16 中记录消耗锌标准溶液的体积(V_1)。

(4) 镁含量的测定。

移取 25.00 mL 试液于小烧杯中，滴加 $NH_3 \cdot H_2O$ 溶液(1 : 1)至刚出现沉淀，再加 HCl 溶液(1 : 1)至沉淀恰好溶解。向其中加入 2 g 固体 NH_4Cl，滴加六次甲基四胺溶液至沉淀出现并过量 15 mL。试液加热至 80℃并保持 5 min，冷却后过滤，用少量蒸馏水洗涤沉淀数次。滤液和洗涤液接入 250 mL 锥形瓶中。再向其中加入 4 mL 三乙醇胺(1 : 2)、10 mL NH_3-NH_4Cl 缓冲溶液(pH ≈ 10)、1 滴甲基红指示剂和 3～5 滴 EBT 指示剂，用 EDTA 标准溶液滴定至溶液由暗红色变为蓝绿色，即为终点。平行滴定 3 次。在表 3.16 中记录消耗 EDTA 标准溶液的体积(V_2)。

数据记录与处理

(1) 计算药片中 Al 的含量[以 $Al(OH)_3$ 表示]。

(2) 计算药片中 Mg 的含量(以 MgO 表示)。

表 3.16　铝和镁含量的测定

	1	2	3
V_1(终)/mL			
V_1(始)/mL			
V_1/mL			
$w_{Al(OH)_3}$ /%			
$\|d_i\|$			
$\bar{d}_r$ /%			
V_2(终)/mL			
V_2(始)/mL			
V_2/mL			
w_{MgO} /%			
$\|d_i\|$			
$\bar{d}_r$ /%			

注意事项

(1)每片胃舒平药片试样中铝和镁的含量可能不相等。为使测定结果具有代表性，本实验取较多样品，研细混匀后再取部分进行分析。

(2)测定镁时，加入一滴甲基红能使终点更敏锐。

思考题

(1)为什么不采用直接滴定法测定铝离子？

(2)测定镁含量时，加入三乙醇胺的作用是什么？

实验 15 洗涤剂中 EDTA 含量的测定

实验目的

(1)掌握硫酸铜溶液的配制和标定方法。

(2)掌握测定洗涤剂中 EDTA 含量的原理和方法。

实验原理

EDTA 是洗涤剂中的重要成分，它可以掩蔽金属离子，增加表面活性剂的活性和生成泡沫的稳定性，有利于提高洗涤剂的清洁效率。将试液用盐酸调节 pH 至 4～5 后，以 1-(2-吡啶偶氮)-2-萘酚(PAN)作指示剂，用硫酸铜标准溶液滴定样品中的 EDTA。PAN 指示剂在该 pH 范围内呈黄色，与金属离子的络合物为红色，因此终点溶液颜色变化为黄绿色[CuY(蓝色)+PAN(黄色)]到紫色[CuY(蓝色)+Cu-PAN(红色)]。

实验用品

仪器：微量滴定管(2 mL)，容量瓶(250 mL)，锥形瓶(250 mL)，移液管(25 mL)，分析天平。

试剂：EDTA 标准溶液(0.02 mol/L)，1-(2-吡啶偶氮)-2-萘酚(PAN)(1 g/L，乙醇溶液)，盐酸溶液(5 mol/L)，$CuSO_4·5H_2O$(s，AR)，乙酸缓冲溶液(pH≈4.65，0.4 mol/L 乙酸溶液和 0.2 mol/L 氢氧化钠溶液等体积混合)，洗涤剂试样。

实验步骤

(1)硫酸铜标准溶液(0.01 mol/L)的配制和标定。

称取 2.5 g 硫酸铜($CuSO_4·5H_2O$)，用蒸馏水溶解并稀释至 1000 mL。配制的硫酸铜溶液用 0.02 mol/L EDTA 标准溶液(详见实验 11)进行标定。方法如下：准确移取 10.00 mL EDTA 标准溶液(0.02 mol/L)于 250 mL 锥形瓶中，加 50 mL 蒸馏水稀释，用 HCl 溶液调节 pH 至 4～5。加 5 mL 乙酸缓冲溶液，加热至 60℃，加 5～6 滴 PAN 指示剂，用硫酸铜标准溶液滴定至溶液由黄色变为紫色，并保持 1 min 不褪色即为终点。在表 3.17 中记录消耗硫酸铜标准溶液的体积(V_1)。平行滴定 3 次。

(2)洗涤剂中 EDTA 含量的测定。

准确称取 5.0～5.5 g 样品于 250 mL 锥形瓶中，加水稀释，用 HCl 溶液调节 pH 至 4～5。加 5 mL 乙酸缓冲溶液，加热至 60℃，加 5～6 滴 PAN 指示剂。在搅拌条件下，用微量滴定管滴加硫酸铜标准溶液至溶液由黄色变成紫色，并保持 1 min 不褪色即为终点。在表 3.18 中记录消耗硫酸铜标准溶液的体积(V_2)。平行滴定 3 次。

表 3.17　硫酸铜标准溶液的标定

	1	2	3
V_1(终)/mL			
V_1(始)/mL			
V_1/mL			
c_{CuSO_4} /(mol/L)			
$\bar{c}_{CuSO_4}$ /(mol/L)			
$\lvert d_i \rvert$			
$\bar{d}_r$ /%			

表 3.18　样品中 EDTA 含量的测定

	1	2	3
V_2(终)/mL			
V_2(始)/mL			
V_2/mL			
w_{EDTA}/%			
$\bar{w}_{EDTA}$/%			
$\lvert d_i \rvert$			
$\bar{d}_r$ /%			

数据记录与处理

(1)计算 $CuSO_4$ 标准溶液的浓度、平均值和相对平均偏差($\bar{d}_r \leqslant 0.2\%$)。

(2)计算洗涤剂样品中 EDTA 的质量分数、平均值和相对平均偏差($\bar{d}_r \leqslant 0.2\%$)。

注意事项

(1)洗涤剂中 EDTA 含量一般较少，应在不断搅拌条件下缓慢滴入，以免过量。

(2)试样的 pH 一定要严格控制，可用 pH 计或精密 pH 试纸调节。

思考题

(1)简述洗涤剂中 EDTA 的测定原理。

(2)测定 EDTA 还有其他方法吗？举例说明。

(3)在不同 pH 条件下，Cu^{2+} 与 EDTA 有何反应？

实验 16　水泥熟料中铁、铝、钙、镁含量的测定

实验目的

(1) 巩固络合滴定法中通过控制酸度测定共存组分的原理。

(2) 了解络合滴定中消除干扰组分的方法。

实验原理

水泥熟料的主要化学成分为 SiO_2(18%～24%)、Fe_2O_3(2.0%～5.5%)、Al_2O_3(4.0%～9.5%)、CaO(60%～67%)和 MgO(< 4.5%)。试样与盐酸作用后，生成硅酸和可溶性的氯化物。硅酸是一种无机酸，在水溶液中绝大部分以溶胶状态存在，其化学式为 $SiO_2 \cdot H_2O$。用浓酸和加热等方法处理后，绝大部分硅酸水溶胶脱水变成水凝胶析出，然后利用沉淀分离的方法把硅酸与水泥中的铁、铝、钙、镁等组分分开。

铁、铝、钙、镁等组分分别以 Fe^{3+}、Al^{3+}、Ca^{2+}、Mg^{2+} 等形式存在于滤完 SiO_2 沉淀的滤液中，它们都能与 EDTA 形成稳定的配合物($\lg K_{FeY} = 25.1$，$\lg K_{AlY} = 16.3$，$\lg K_{CaY} = 10.69$，$\lg K_{MgY} = 8.7$)，其稳定性有较显著的差别，可通过控制酸度，用 EDTA 标准溶液分别滴定，测定其含量。

(1) 铁的测定。

在 pH 为 1.5～2.5、温度为 60～70℃的溶液中，以磺基水杨酸或其钠盐为指示剂，用 EDTA 标准溶液滴定铁，终点时溶液由紫红色变为亮黄色。

显色反应　$Fe^{3+} + HIn^- \rightleftharpoons FeIn^+$(紫红色)$+H^+$

滴定反应　$Fe^{3+} + H_2Y^{2-} \rightleftharpoons FeY^- + 2H^+$

终点时　$FeIn^+$(紫红色)$+H_2Y^{2-} \rightleftharpoons FeY^-$(亮黄色)$+HIn^- + H^+$

(2) 铝的测定。

Al^{3+}与 EDTA 的配位反应很慢，不宜采用直接滴定法。在 pH 为 4.3 的条件下，先加入过量的 EDTA 标准溶液，加热煮沸，使 Al^{3+}与 EDTA 充分反应。然后以 PAN 为指示剂，用铜盐标准溶液返滴定。滴定开始前溶液呈黄色(指示剂颜色)，逐渐由黄色变为绿色(PAN 和 CuY 的混合色)，终点时溶液呈紫色(PAN 和 CuPAN 的混合色)。

滴定反应　$Al^{3+} + H_2Y^{2-}$(过量)$\rightleftharpoons AlY^-$(无色)$+2H^+$

铜盐返滴定过量 EDTA　$Cu^{2+} + H_2Y^{2-}$(剩余)$\rightleftharpoons CuY^{2-}$(蓝色)$+2H^+$

终点时　Cu^{2+} + PAN(黄色)$\rightleftharpoons$ Cu-PAN(红色)

溶液中蓝色 CuY 量的多少对终点颜色变化的敏锐程度有影响。因此，过量 EDTA 的量要加以控制。一般 100 mL 溶液中加入的 EDTA 标准溶液(0.01～0.015 mol/L)以过量 10～15 mL 为宜。

(3) 钙的测定。

在 pH≥12 的强碱性溶液中，Mg^{2+}形成 $Mg(OH)_2$ 沉淀而被掩蔽，Fe^{3+}、Al^{3+}用三乙醇胺掩蔽。以钙黄绿素-甲基百里香酚蓝-酚酞(CMP)为混合指示剂，用 EDTA 标准溶液滴定钙。

pH > 12 时，钙黄绿素本身呈橘红色，与 Ca^{2+}、Sr^{2+}、Ba^{2+}等离子配位后呈绿色的荧光。终点时，溶液呈橘红色，但由于溶液中有残余荧光，会影响终点的观察。本实验中，混合指示剂中的甲基百里香酚蓝和酚酞在滴定中可掩盖钙黄绿素的残余荧光。

(4) 镁的测定。

以 EDTA 络合滴定法测定镁的含量多采用差减法，即在一份溶液中，调节 $pH \approx 10$，用 EDTA 滴定钙、镁总含量，从总含量中减去钙的量，即求得镁的含量。

滴定钙、镁总含量时，常用的指示剂有铬黑 T 和酸性铬蓝 K-萘酚绿 B (K-B) 混合指示剂。铬黑 T 易被某些重金属离子封闭，所以采用 K-B 混合指示剂作为 EDTA 滴定钙、镁总含量的指示剂。萘酚绿 B 在滴定过程中没有颜色变化，只起衬托终点颜色的作用，终点颜色的变化是红色到蓝色。Fe^{3+}、Al^{3+}用三乙醇胺和酒石酸钾钠进行联合掩蔽。

实验用品

仪器：烧杯 (100 mL，1000 mL)，量筒 (100 mL)，移液管 (5 mL，10 mL，25 mL)，容量瓶 (250 mL)，锥形瓶 (250 mL)，酸式滴定管 (50 mL)，分析天平。

试剂：EDTA 标准溶液 (0.01 mol/L，见实验 11)，$CuSO_4$ 标准溶液 (0.01 mol/L，见实验 15)，磺基水杨酸 (10%)，溴甲酚绿指示剂 (0.05%)，PAN 指示剂 (0.3%)，K-B 指示剂，CMP 指示剂 (准确称取 1 g 钙黄绿素、1 g 甲基百里香酚蓝、0.2 g 酚酞，与 50 g 105℃烘干的硝酸钾混合研细，保存于磨口瓶中)，HAc-NaAc 缓冲溶液 ($pH \approx 4.3$，将 33.7 g 无水乙酸钠溶于水中，加入 80 mL 冰醋酸，加水稀释至 1 L，摇匀)，酒石酸钾钠 (10%，取 10 g 酒石酸钾钠溶于 90 mL 水中)，NH_3-NH_4Cl 缓冲溶液 ($pH \approx 10$)，HCl 溶液 (6 mol/L)，浓 HCl，浓 HNO_3，KOH 溶液 (20%)，氨水 (1∶1)，三乙醇胺溶液 (1∶2)。

实验步骤

准确称取 0.5 g 试样于 100 mL 烧杯中，沿烧杯口滴加 10 mL 浓 HCl 及 1 mL 浓 HNO_3，仔细搅拌，使所有深灰色试样变为淡黄色糊状物。盖上表面皿，将烧杯放在电热板 (或沸水浴) 上加热 (通风橱内)，使可溶性的盐类溶解。过滤，滤液置于 250 mL 容量瓶中，用滴管以热盐酸 (3∶97) 擦洗玻璃棒和烧杯，并洗涤残渣 3～4 次，然后用热水充分洗涤残渣 (约 10 次，直至检验无氯离子为止)。滤液和洗涤液合并于 250 mL 容量瓶中，冷却至室温，加蒸馏水至刻度，摇匀，待用。

(1) Fe_2O_3 的测定。

准确量取 25.00 mL 试样溶液于 250 mL 锥形瓶中，加 50 mL 水和 2 滴 0.05%溴甲酚绿指示剂 ($pH < 3.8$ 时呈黄色，$pH > 5.4$ 时呈绿色)，逐滴加入氨水 (1∶1)，使其呈绿色，然后用 HCl 溶液 (6 mol/L) 调至黄色后再过量 3 滴，溶液 pH 为 1.8～2.0。将溶液加热至 60～70℃，加 10 滴磺基水杨酸指示剂，用 EDTA 标准溶液缓慢滴定至溶液由紫红色变为亮黄色 (终点时溶液温度不低于 60℃，保留此溶液供测定 Al_2O_3 含量用)。在表 3.19 中记录消耗的 EDTA 标准溶液的体积 (V_{EDTA})。

(2) Al_2O_3 的测定。

在滴定完铁后的溶液中准确加入 25.00 mL EDTA 标准溶液 (0.01 mol/L)，加 15 mL HAc-NaAc 缓冲溶液 ($pH \approx 4.3$)。煮沸 1～2 min，稍冷，加 4～5 滴 PAN 指示剂，用 $CuSO_4$ 标准溶液 (0.01 mol/L) 滴定至亮紫色。在表 3.20 中记录消耗的 $CuSO_4$ 标准溶液的体积 (V_{CuSO_4})。

(3) CaO 的测定。

准确移取 10.00 mL 试样溶液于 250 mL 锥形瓶中，加 50 mL 蒸馏水、5 mL 三乙醇胺溶液(1 ∶ 2)及少许 CMP 指示剂，在搅拌下加入 20% KOH 溶液至出现绿色荧光，再过量 5～8 mL(溶液 pH > 13)。用 EDTA 标准溶液滴定至溶液呈红色，即为终点。在表 3.21 中记录消耗的 EDTA 标准溶液的体积(V_1)。

(4) MgO 的测定。

准确移取 10.00 mL 试样溶液于 250 mL 锥形瓶中，加 50 mL 蒸馏水、1 mL 酒石酸钾钠溶液和 5 mL 三乙醇胺溶液(1 : 2)，振摇 1 min。然后加入 15 mL NH_3-NH_4Cl 缓冲溶液(pH ≈ 10)及少许 K-B 指示剂。用 EDTA 标准溶液(0.01 mol/L)滴定至溶液由紫红色变为蓝色。在表 3.22 中记录消耗的 EDTA 标准溶液的体积(V_2)。

数据记录与处理

分别计算水泥样品中铁、铝、钙、镁的含量。

$$w_{Fe_2O_3}=\frac{c_{EDTA}\times V_{EDTA}\times M_{Fe_2O_3}\times 250}{2m_s\times 1000\times 25.00}\times 100\%$$

$$w_{Al_2O_3}=\frac{(c_{EDTA}\times 25.00-c_{CuSO_4}\times V_{CuSO_4})\times M_{Al_2O_3}\times 250}{2m_s\times 1000\times 25.00}\times 100\%$$

$$w_{CaO}=\frac{c_{EDTA}\times V_1\times M_{CaO}\times 250}{m_s\times 1000\times 10.00}\times 100\%$$

$$w_{MgO}=\frac{c_{EDTA}\times (V_2-V_1)\times M_{MgO}\times 250}{m_s\times 1000\times 10.00}\times 100\%$$

表 3.19 Fe_2O_3 含量的测定

	1	2	3
V_{EDTA}(终)/mL			
V_{EDTA}(始)/mL			
V_{EDTA}/mL			
$w_{Fe_2O_3}$ /%			
$\overline{w}_{Fe_2O_3}$ /%			
$\lvert d_i \rvert$			
$\overline{d}_r$ /%			

表 3.20 Al_2O_3 含量的测定

	1	2	3
V_{CuSO_4}(终)/mL			
V_{CuSO_4}(始)/mL			
V_{CuSO_4} /mL			
$w_{Al_2O_3}$ /%			
$\overline{w}_{Al_2O_3}$ /%			
$\lvert d_i \rvert$			
$\overline{d}_r$ /%			

表 3.21　CaO 含量的测定

	1	2	3
V_1(终)/mL			
V_1(始)/mL			
V_1/mL			
w_{CaO}/%			
$\bar{w}_{CaO}$/%			
$\|d_i\|$			
$\bar{d}_r$/%			

表 3.22　MgO 含量的测定

	1	2	3
V_2(终)/mL			
V_2(始)/mL			
V_2/mL			
w_{MgO}/%			
$\bar{w}_{MgO}$/%			
$\|d_i\|$			
$\bar{d}_r$/%			

思考题

(1) 滴定 Fe^{3+}时，如何消除 Al^{3+}、Ca^{2+}、Mg^{2+}等离子的干扰？

(2) 如果 Fe^{3+}的测定结果不准确，对 Al^{3+}的测定结果有什么影响？

(3) EDTA 滴定 Al^{3+}时，为什么要采用返滴定法？

(4) 测定 Ca^{2+}、Mg^{2+}时加入三乙醇胺的目的是什么？测定 Ca^{2+}时为什么要在加入 KOH 之前加三乙醇胺？

实验 17　工业硫酸铝中铝含量的测定

实验目的

(1) 了解返滴定法的基本原理。

(2) 掌握置换滴定法测定铝含量的原理和方法。

实验原理

Al^{3+} 容易形成多核羟基络合物，如$[Al_2(H_2O)_6(OH)_3]^{3+}$、$[Al_2(H_2O)(OH)_6]^{3+}$等，因此 Al^{3+}与 EDTA 的络合速度缓慢，需加入过量 EDTA 并加热煮沸，络合反应才会进行得比较完全，所以宜采用返滴定法或置换滴定法。

返滴定法是加入定量且过量的 EDTA 标准溶液，调节 pH ≈ 3.5，煮沸数分钟，使 Al^{3+}与 EDTA 反应完全，然后调节溶液 pH 为 5～6，用 $CuSO_4$ 标准溶液返滴定过量的 EDTA。

置换滴定法是加入过量的 NH_4F，加热煮沸，利用 F^- 与 Al^{3+} 生成更稳定络合物的性质，置换出与 Al^{3+} 物质的量相等的 EDTA，再用 $CuSO_4$ 标准溶液滴定释放出来的 EDTA。

设 EDTA 标准溶液的浓度为 c_{EDTA} (mol/L)，标定 $CuSO_4$ 标准溶液时消耗的体积为 V_1(mL)，滴定样品时消耗 $CuSO_4$ 标准溶液的体积为 V_2(mL)，样品质量为 m(g)，则样品中铝的含量可用下列公式计算：

$$w_{Al}=\frac{26.98\times c_{EDTA}\times V_2\times 25.00\times 250}{V_1\times m\times 25.00}\times 100\%$$

实验用品

仪器：酸式滴定管(50 mL)，锥形瓶(250 mL)，移液管(25 mL)，烧杯(1000 mL)，分析天平。

试剂：EDTA 标准溶液(0.02 mol/L)，$CuSO_4\cdot 5H_2O$，PAN 指示剂(0.1%，0.1 g PAN 溶于 100 mL 无水乙醇)，百里酚蓝指示剂(0.1%，0.1 g 百里酚蓝溶于 100 mL 20%乙醇)，HCl (1 ∶ 1)，H_2SO_4 (1 ∶ 1)，$NH_3\cdot H_2O$(1 ∶ 1)，NH_4F，六次甲基四胺缓冲溶液(20%)，工业硫酸铝。

实验步骤

(1) EDTA 标准溶液(0.02 mol/L)的配制和标定。

详见实验 11。

(2) $CuSO_4$ 标准溶液(0.02 mol/L)的配制和标定。

称取 2.6 g $CuSO_4\cdot 5H_2O$，加入 2～3 滴 H_2SO_4(1 ∶ 1)，加 500 mL 水溶解，摇匀。准确移取 25.00 mL EDTA 标准溶液于 250 mL 锥形瓶中，加入 10 mL 20%六次甲基四胺缓冲溶液，加热至 80～90℃后，加入 2～3 mL 0.1% PAN 指示剂，用 $CuSO_4$ 标准溶液滴定至呈稳定的紫红色。在表 3.23 中记录消耗 $CuSO_4$ 标准溶液的体积(V_1)。平行滴定 3 次。

(3) 铝的测定。

准确称取 1.3 g 工业硫酸铝于 100 mL 烧杯中，加入 10 mL HCl 溶液(1 ∶ 1)，用 50 mL 蒸馏水溶解后转移至 250 mL 容量瓶中，用蒸馏水稀释至刻度，摇匀。

准确移取 25.00 mL 试液于 250 mL 锥形瓶中，加入 30 mL EDTA 标准溶液及 5 滴百里酚蓝指示剂，再滴加 $NH_3\cdot H_2O$(1 ∶ 1)至溶液恰呈黄色(pH≈3)。煮沸 2 min，加入 10 mL 六次甲基四胺缓冲溶液(20%)和 2～3 mL PAN 指示剂，趁热用 $CuSO_4$ 标准溶液滴定至溶液呈稳定的紫红色，不用记录 $CuSO_4$ 标准溶液的用量。于滴定后的溶液中加入 1～2 g NH_4F，加热煮沸 2 min(必要时补加 8 滴 PAN 指示剂)，再用 $CuSO_4$ 标准溶液滴定至紫红色。在表 3.24 中记录消耗 $CuSO_4$ 标准溶液的体积(V_2)。平行滴定 3 次。

数据记录与处理

计算 EDTA 标准溶液和 $CuSO_4$ 标准溶液的浓度，以及样品中铝的质量分数。

表 3.23　硫酸铜标准溶液的标定

	1	2	3
V_1(终)/mL			
V_1(始)/mL			
V_1/mL			
c_{CuSO_4} /(mol/L)			
$\bar{c}_{CuSO_4}$ /(mol/L)			
$\|d_i\|$			
$\bar{d}_r$ /%			

表 3.24　铝含量的测定

	1	2	3
V_2(终)/mL			
V_2(始)/mL			
V_2/mL			
w_{Al}/%			
$\bar{w}_{Al}$/%			
$\|d_i\|$			
$\bar{d}_r$ /%			

注意事项

(1) 加热时要用玻璃棒搅拌以防止溶液暴沸溅出。

(2) 指示剂的用量要适中。

思考题

(1) 标定 $CuSO_4$ 标准溶液时，在加入六次甲基四胺缓冲溶液后，为什么要将溶液加热至 80～90℃？

(2) 测定铝含量时需加 30 mL EDTA 溶液，是否必须准确加入？为什么？

(3) 加入 NH_4F 的目的是什么？NH_4F 的量太多或太少对测定各有什么影响？

3.4　氧化还原滴定实验

实验 18　高锰酸钾标准溶液的配制、标定和 H_2O_2 含量的测定

实验目的

(1) 掌握高锰酸钾标准溶液的配制和标定方法。

(2) 掌握应用高锰酸钾法测定过氧化氢含量的原理和方法。

实验原理

标定 $KMnO_4$ 标准溶液的基准物质有 As_2O_3、纯铁丝和 $Na_2C_2O_4$ 等，其中以 $Na_2C_2O_4$ 最常用。在酸性介质中，$KMnO_4$ 与 $Na_2C_2O_4$ 发生如下反应：

$$2MnO_4^- + 5C_2O_4^{2-} + 16H^+ = 2Mn^{2+} + 10CO_2\uparrow + 8H_2O$$

因溶液本身有紫红色，不需外加指示剂。

H_2O_2 又称双氧水，在工业、生物、医药等方面应用广泛。可用于漂白毛、丝织物及消毒、杀菌；纯 H_2O_2 可作为火箭燃料的氧化剂；工业上可利用 H_2O_2 的还原性除去氯气；生物方面，可以利用过氧化氢酶对 H_2O_2 分解反应的催化作用，测量过氧化氢酶的活性。

在室温条件下，H_2O_2 在稀 H_2SO_4 溶液中能定量地被 $KMnO_4$ 溶液氧化，因此可以用 $KMnO_4$ 法测定过氧化氢的含量。

$$2MnO_4^- + 5H_2O_2 + 6H^+ = 2Mn^{2+} + 5O_2\uparrow + 8H_2O$$

实验用品

仪器：酸式滴定管(50 mL)，锥形瓶(250 mL)，烧杯(1000 mL)，量筒(10 mL，50 mL)，容量瓶(250 mL)，移液管(25 mL)，分析天平。

试剂：$Na_2C_2O_4$(基准物质)，$KMnO_4$ 标准溶液(0.02 mol/L)，H_2SO_4 溶液(3 mol/L)，$MnSO_4$ 溶液(1 mol/L)，$KMnO_4$(AR)，H_2O_2 试样(市售质量分数约为30%的 H_2O_2 水溶液)。

实验步骤

(1) $KMnO_4$ 标准溶液(0.02 mol/L)的配制。

称取 1.6 g $KMnO_4$ 溶于 500 mL 蒸馏水中，盖上表面皿，加热煮沸 20～30 min(中途补加一定量的蒸馏水以保持溶液体积基本不变)。冷却后将溶液转移至棕色瓶内，在暗处放置 7～10 天(如果溶液经煮沸并在水浴上保温 1 h，放置 2～3 天也可)，然后用微孔玻璃漏斗(3 号或 4 号)过滤除去 MnO_2 等杂质。滤液储存于洁净的带塞棕色瓶中，待标定。

(2) $KMnO_4$ 标准溶液(0.02 mol/L)的标定。

准确称取 0.13～0.16 g $Na_2C_2O_4$ 基准物质于 250 mL 锥形瓶中，加 40 mL 水、10 mL H_2SO_4(3 mol/L)。加热至 70～80℃(开始冒蒸汽时的温度)，趁热用 $KMnO_4$ 溶液进行滴定。临近终点时滴定速度要减慢，同时充分摇匀，直到溶液呈现微红色并持续 30 s 不褪色即为终点。注意终点时溶液的温度应保持 60℃以上。平行滴定 3 次。将消耗的 $KMnO_4$ 标准溶液的体积记录在表 3.25 中。

(3) H_2O_2 含量的测定。

移取 2.00 mL H_2O_2 试样溶液于 250 mL 容量瓶中，加水稀释至刻度，摇匀。移取 25.00 mL 稀释液置于 250 mL 锥形瓶中，加入 5 mL H_2SO_4 溶液(3 mol/L)及 2～3 滴 $MnSO_4$ 溶液(1 mol/L)，用 $KMnO_4$ 标准溶液滴定至溶液呈微红色，且 30 s 不褪色即为终点。平行滴定 3 次。将消耗的 $KMnO_4$ 标准溶液的体积记录在表 3.26 中。

表 3.25　$KMnO_4$ 标准溶液的标定

	1	2	3
$m_{Na_2C_2O_4}$ /g			
V_{KMnO_4} (终)/mL			
V_{KMnO_4} (始)/mL			
V_{KMnO_4} /mL			
c_{KMnO_4} /(mol/L)			
$\bar{c}_{KMnO_4}$ /(mol/L)			
$\vert d_i \vert$			
$\bar{d}_r$ /%			

表 3.26　H_2O_2 含量的测定

	1	2	3
$V_{H_2O_2}$ /mL			
V_{KMnO_4} (终)/mL			
V_{KMnO_4} (始)/mL			
V_{KMnO_4} /mL			
$\rho_{H_2O_2}$ /(g/L)			
$\bar{\rho}_{H_2O_2}$ /(g/L)			
$\vert d_i \vert$			
$\bar{d}_r$ /%			

数据记录与处理

(1) 根据标定时消耗的 $KMnO_4$ 标准溶液的体积，计算 $KMnO_4$ 标准溶液的浓度和相对平均偏差。

(2) 根据 $KMnO_4$ 标准溶液的浓度和滴定时所消耗的体积以及滴定前样品的稀释情况，计算样品中 H_2O_2 的含量(g/L)和相对平均偏差。

注意事项

(1) 室温下，高锰酸钾与草酸钠之间的反应速度缓慢，故必须将溶液加热。但温度不能太高，若超过 90℃，易引起草酸分解：

$$H_2C_2O_4 = CO_2\uparrow + CO\uparrow + H_2O$$

(2) 如滴定速度过快，部分高锰酸钾将来不及与草酸钠反应，而在热的酸性溶液中分解：

$$4MnO_4^- + 4H^+ = 4MnO_2 + 3O_2\uparrow + 2H_2O$$

思考题

(1) 用 $Na_2C_2O_4$ 为基准物质标定 $KMnO_4$ 标准溶液时应在什么条件下进行？

(2) 标定 $KMnO_4$ 标准溶液时，为什么第一滴 $KMnO_4$ 加入后溶液的红色褪去很慢，而以后红色褪去的速度越来越快？

(3) 用高锰酸钾法测定 H_2O_2 时，能否用 HNO_3 或 HCl 溶液调节酸度？为什么？

实验 19　重铬酸钾法测定铁矿石中的铁含量(无汞法)

实验目的

(1) 掌握无汞法测定铁的原理和方法。

(2) 了解无汞法测定铁的绿色环保意义。

实验原理

铁矿石的种类很多，用于炼铁的主要有磁铁矿(Fe_3O_4)、赤铁矿(Fe_2O_3)和菱铁矿($FeCO_3$)等。试样用 HCl 溶解后，其中的铁转化为 Fe^{3+}，首先在热的浓 HCl 溶液中用 $SnCl_2$ 将大部分 Fe(Ⅲ)还原为 Fe(Ⅱ)，再用 $TiCl_3$ 还原剩余的 Fe(Ⅲ)，反应方程式为

$$2Fe^{3+} + SnCl_4^{2-} + 2Cl^- = 2Fe^{2+} + SnCl_6^{2-}$$

$$Fe^{3+} + Ti^{3+} + H_2O = Fe^{2+} + TiO^{2+} + 2H^+$$

当全部 Fe(Ⅲ)定量还原为 Fe(Ⅱ)后，稍过量的 $TiCl_3$ 即可将溶液中的预处理指示剂 Na_2WO_4 由无色还原为蓝色的 W(Ⅴ)(俗称钨蓝)。然后用少量的稀 $K_2Cr_2O_7$ 溶液将过量的钨蓝氧化，使蓝色恰好消失，从而指示预还原的终点。

预处理后，在硫磷混酸介质中，以二苯胺磺酸钠为指示剂，用 $K_2Cr_2O_7$ 标准溶液滴定至溶液呈紫色，即达终点。

$SnCl_2$-$TiCl_3$-$K_2Cr_2O_7$ 无汞法测铁避免了有汞法对环境的污染，目前已被定为铁矿石分析的国家标准。

实验用品

仪器：酸式滴定管(50 mL)，锥形瓶(250 mL)，量筒(100 mL)，容量瓶(250 mL)，移液管(10 mL，25mL)，烧杯(100 mL)，表面皿，分析天平等。

试剂：$K_2Cr_2O_7$ 标准溶液(0.017 mol/L)，浓 HCl，$SnCl_2$ 溶液(50 g/L)，$TiCl_3$ 溶液(15 g/L)，硫磷混酸溶液，Na_2WO_4 溶液(250 g/L)，二苯胺磺酸钠水溶液(2 g/L)，铁矿石试样。

实验步骤

(1) $K_2Cr_2O_7$ 标准溶液(0.017 mol/L)的配制。

准确称取 1.2～1.3 g $K_2Cr_2O_7$ 于 100 mL 烧杯中，加适量水溶解后定量转入 250 mL 容量瓶中，用蒸馏水稀释至刻度，摇匀。计算 $K_2Cr_2O_7$ 标准溶液的浓度。

(2) 试样的溶解。

称取约 0.2 g 铁矿石试样置于 250 mL 锥形瓶中，用少量水润湿，加入 10 mL 浓 HCl，并加 8～10 滴 $SnCl_2$ 溶液助溶。盖上表面皿，在近沸的水浴中加热 20～30 min，至残渣变为白色，表明试样溶解完全，此时溶液呈棕黄色。用少量蒸馏水冲洗表面皿和锥形瓶内壁。

(3) 试样的预处理。

趁热用滴管向试液中小心滴加 $SnCl_2$ 溶液还原 Fe(III)，边滴边摇，直到溶液由棕黄色变为浅黄色，表明大部分 Fe(III) 已被还原。加入 4 滴 Na_2WO_4 和 60 mL 水，加热。在摇动下逐滴加入 $TiCl_3$ 至溶液出现稳定的浅蓝色。用自来水冲洗锥形瓶外壁使溶液冷却至室温，然后小心滴加稀释 10 倍的 $K_2Cr_2O_7$ 溶液，至蓝色刚好消失。

(4) 铁的测定。

将试液加蒸馏水稀释至 150 mL，加入 15 mL 硫磷混酸和 5～6 滴二苯胺磺酸钠指示剂，立即用 $K_2Cr_2O_7$ 标准溶液测定，至溶液呈稳定的紫色即为终点。在表 3.27 中记录消耗 $K_2Cr_2O_7$ 标准溶液的体积。平行滴定 3 次。

数据记录与处理

根据消耗的重铬酸钾标准溶液的体积，计算铁矿石中铁的质量分数和相对平均偏差。

表 3.27　铁含量的测定

	1	2	3
$m_{K_2Cr_2O_7}$ /g			
$c_{K_2Cr_2O_7}$ /(mol/L)			
$V_{K_2Cr_2O_7}$ (终)/mL			
$V_{K_2Cr_2O_7}$ (始)/mL			
$V_{K_2Cr_2O_7}$ /mL			
c_{Fe}/(mol/L)			
w_{Fe}/%			
$\bar{w}_{Fe}$ /%			
$\|d_i\|$			
$\bar{d}_r$ /%			

思考题

(1) 样品预处理时，为什么 $SnCl_2$ 溶液要趁热逐滴加入？

(2) 预还原 Fe(III) 至 Fe(Ⅱ) 时，为什么要用 $SnCl_2$ 和 $TiCl_3$ 两种还原剂？只使用一种还原剂有什么缺点？

(3) 滴定前加入 H_3PO_4 的作用是什么？加入 H_3PO_4 后为什么要立即滴定？

实验 20　环境水样中化学需氧量的测定($KMnO_4$ 法)

实验目的

(1) 初步了解环境分析的重要性及水样的采集和保存方法。

(2) 了解化学需氧量(COD)的含义、表示方法及其测定的意义。

(3) 掌握高锰酸钾法测定水中 COD 的原理及方法。

实验原理

化学需氧量(chemical oxygen demond，COD)是水质监测的一项重要指标。它是指氧化一定体积的水样中还原性物质所需消耗的强氧化剂的量，以对应的 O_2 量(mg/L)表示。废水中大部分的还原性物质是有机物，因此 COD 是评价有机物导致的水污染的重要指标。COD 的测定可采用高锰酸钾法和重铬酸钾法。高锰酸钾法适用于地表水、饮用水和生活污水等污染不十分严重的水体；而重铬酸钾法适用于工业废水。

本实验采用高锰酸钾法测定水样的 COD。在酸性条件下，向被测水样中定量加入高锰酸钾标准溶液，加热使高锰酸钾与水样中的有机污染物充分反应。过量的高锰酸钾用一定量的草酸钠标准溶液还原，再用高锰酸钾标准溶液返滴定过量的草酸钠。根据高锰酸钾标准溶液的消耗量，计算水样的化学需氧量。主要反应方程式如下：

$$4MnO_4^-(\text{过量})+5C+12H^+ = 4Mn^{2+}+5CO_2\uparrow+6H_2O$$

$$2MnO_4^-(\text{剩余量})+5C_2O_4^{2-}(\text{过量})+16H^+ = 2Mn^{2+}+10CO_2\uparrow+8H_2O$$

$$5C_2O_4^{2-}(\text{剩余量})+2MnO_4^-(\text{滴定剂})+16H^+ = 2Mn^{2+}+10CO_2\uparrow+8H_2O$$

这里，C 泛指水中的还原性物质或需氧物质，主要为有机物。

实验用品

仪器：酸式滴定管(50 mL)，容量瓶(250 mL)，锥形瓶(250 mL)，烧杯(100 mL)，分析天平等。

试剂：$KMnO_4$ 溶液(0.02 mol/L，0.002 mol/L)，$Na_2C_2O_4$(s)，H_2SO_4 溶液(6 mol/L)，硫酸银(s)。

实验步骤

(1) $Na_2C_2O_4$ 标准溶液(0.005 mol/L)的配制。

准确称取约 0.17 g $Na_2C_2O_4$(100～105℃干燥 2 h)于 100 mL 烧杯中，加蒸馏水溶解后定量转移至 250 mL 容量瓶中，稀释、定容、摇匀。

(2) $KMnO_4$ 标准溶液(0.002 mol/L)的配制和标定。

准确移取 25.00 mL 0.02 mol/L $KMnO_4$ 溶液于 250 mL 容量瓶中，加水稀释至刻度，摇匀，置于暗处。准确移取 25.00 mL $Na_2C_2O_4$ 标准溶液于 250 mL 锥形瓶中，加入 5 mL H_2SO_4 溶液(6 mol/L)，在水浴上加热到 75～85℃，趁热用 $KMnO_4$ 标准溶液(0.002 mol/L)滴定(滴定速度由慢到快，再到慢)至溶液呈微红色并保持 30 s 不褪色。在表 3.28 中记录数据，平行滴定 3 次。

(3) 水样中 COD 的测定。

准确移取 100.00 mL 水样于 250 mL 锥形瓶中，加 5 mL H_2SO_4 溶液(6 mol/L)和几粒沸石，由滴定管准确加入 10.00 mL $KMnO_4$ 标准溶液(0.002 mol/L)，立即加热煮沸，从冒出第一个大气泡开始用小火煮沸 10 min(红色不应褪去，若褪去，应补加 0.002 mol/L $KMnO_4$ 标准溶液至样品呈稳定的红色)，加入的 $KMnO_4$ 标准溶液的总体积(V_1)记录于表 3.29。取下锥形瓶冷却 1 min，准确加入 10.00 mL $Na_2C_2O_4$ 标准溶液，充分摇匀(此时溶液由红色变为无色，否则应增加 $Na_2C_2O_4$ 标准溶液的用量)，趁热用 $KMnO_4$ 标准溶液(0.002 mol/L)滴定至微红色并保持 30 s 不褪色。将 $KMnO_4$ 标准溶液消耗量(V_2)记录在表 3.29 中，平行滴定 3 次。

(4)空白样 COD 的测定。

另取 100.00 mL 蒸馏水代替水样进行实验，操作同(3)，求得空白值。

表 3.28　$KMnO_4$ 标准溶液的标定

	1	2	3
V_{KMnO_4}(终)/mL			
V_{KMnO_4}(始)/mL			
V_{KMnO_4}/mL			
c_{KMnO_4}/(mol/L)			
$\bar{c}_{KMnO_4}$/(mol/L)			
$\lvert d_i \rvert$			
$\bar{d}_r$/%			

表 3.29　水样中 COD 的测定

	1	2	3
V_1/mL			
V_2/mL			
(V_1+V_2)/mL			
$COD_{总}$/(mg/L)			
$\overline{COD}_{总}$/(mg/L)			
$\lvert d_i \rvert$			
$\bar{d}_r$/%			
$COD_{空白}$/(mg/L)			
$\overline{COD}_{空白}$/(mg/L)			
COD_{Mn}/(mg/L)			

数据记录与处理

(1)计算 $KMnO_4$ 标准溶液的浓度、平均值和相对平均偏差($\bar{d}_r \leqslant 0.2\%$)。

(2)计算水样的化学需氧量(COD_{Mn})。

$$COD/(mg/L)=\frac{\left[\frac{5}{4}c_{KMnO_4}\times(V_1+V_2)-\frac{1}{2}c_{Na_2C_2O_4}\times V_{Na_2C_2O_4}\right]\times M_{O_2}\times 1000}{V_{水样}}$$

$$COD_{Mn}=\overline{COD}_{总}-\overline{COD}_{空白}$$

注意事项

(1)水中含氯离子含量超过 300 mg/L 时，将影响测定结果。加水稀释，降低氯离子浓度，可消除干扰。若还不能消除干扰，可加入适量硫酸银。

(2)取样后应及时分析测定。如需放置，可加少量硫酸使 pH < 2，以抑制微生物对有机物的分解，0～5℃保存，并在 48 h 内测定。

(3)分析测定时需加热至沸，此时溶液仍应保持高锰酸钾的紫红色。若红色消失，说明水中有机物较多，应补加适量的高锰酸钾标准溶液。

(4)水样量的多少可初步判断：洁净透明的水样取 100 mL；污染严重、浑浊的水样取 10～30 mL，用蒸馏水稀释至 100 mL。

思考题

(1)水样中氯离子含量高时对测定有何干扰？应采用什么方法消除？

(2)水样中加入高锰酸钾溶液并在沸水中加热后应是什么颜色？若无色说明什么问题？如何处理？

(3)本实验为什么采用返滴定法？

(4)哪些因素会影响 COD 的测定结果？为什么？

实验 21　污水中化学需氧量的测定($K_2Cr_2O_7$ 法)

实验目的

(1)了解测定 COD 的意义和方法。

(2)掌握重铬酸钾法测定 COD 的原理和方法。

实验原理

化学需氧量(COD)是表征水中还原性物质的一个指标，它能反映水体被有机物污染的状况。重铬酸钾法测定水样中 COD 的原理是：在强酸性溶液中，准确加入过量重铬酸钾标准溶液，加热回流，将水样中的还原性物质氧化，过量的重铬酸钾以邻菲咯啉作指示剂，用硫酸亚铁铵标准溶液回滴，根据消耗的重铬酸钾标准溶液计算出水样的 COD。反应式如下：

$$6Fe(NH_4)_2(SO_4)_2+K_2Cr_2O_7+7H_2SO_4 = 3Fe_2(SO_4)_3+Cr_2(SO_4)_3+K_2SO_4+6(NH_4)_2SO_4+7H_2O$$

实验中可加入硫酸银作催化剂，酸性重铬酸钾能氧化包括直链脂肪族化合物在内的大部分有机化合物。氯离子能被重铬酸盐氧化，并且能与硫酸银作用产生沉淀，影响测定结果，故在回流前向水样中加入硫酸汞，将氯离子络合以消除干扰。氯离子含量高于 1000 mg/L 的样品应先稀释再测定。

实验用品

仪器：酸式滴定管(50 mL)，移液管(10 mL)，带锥形瓶(250 mL)的回流装置，烧杯(250 mL)，分析天平，电炉等。

试剂：试亚铁灵指示液[称取 1.485 g 邻菲咯啉($C_{12}H_8N_2 \cdot H_2O$)、0.695 g $FeSO_4 \cdot 7H_2O$ 溶于水中，稀释至 100 mL，储于棕色瓶中]，重铬酸钾标准溶液(0.25 mol/L)，硫酸亚铁铵标准溶液(0.1 mol/L)，硫酸汞(s)，硫酸银(s)，硫酸溶液(6 mol/L)，水样。

实验步骤

(1) 重铬酸钾标准溶液(0.25 mol/L)的配制。

称取约 18.4 g 重铬酸钾(120℃烘干 2 h)溶于水中，定量转移至 250 mL 容量瓶中，稀释至刻度，摇匀。

(2) 硫酸亚铁铵标准溶液(0.1 mol/L)配制和标定。

称取约 9.8 硫酸亚铁铵于 250 mL 烧杯中，加少量水溶解，边搅拌边缓慢加入 20 mL 硫酸(6 mol/L)，溶解后加水稀释至 250 mL，储于试剂瓶中。

准确移取 10.00 mL 重铬酸钾标准溶液于 250 mL 锥形瓶中，加 30 mL 水，缓慢加入 30 mL 硫酸(6 mol/L)，混匀，冷却后加入 3 滴试亚铁灵指示液，用硫酸亚铁铵标准溶液滴定，溶液由黄色经蓝绿色至红褐色为终点。在表 3.30 中记录消耗的硫酸亚铁铵标准溶液的体积。平行滴定 3 次。

(3) 水样中 COD 的测定。

准确移取 20.00 mL 混合均匀的水样(或适量水样稀释至 20.00 mL)置于 250 mL 锥形瓶中，加入 0.4 g 硫酸汞，准确加入 10.00 mL 重铬酸钾标准溶液(0.25 mol/L)及数粒沸石，连接回流冷凝管，从冷凝管上口慢慢加入 30 mL 硫酸-硫酸银溶液，混匀后，加热回流 2 h(自开始沸腾计时)。冷却后，用 90 mL 蒸馏水洗涤冷凝管壁，使溶液总体积不少于 140 mL，取下锥形瓶。溶液再度冷却后，加三滴试亚铁灵指示液，用硫酸亚铁铵标准溶液滴定，溶液由黄色经蓝绿色至红褐色即为终点。在表 3.31 中记录硫酸亚铁铵标准溶液的体积。平行滴定 3 次。

(4) 空白样中 COD 的测定。

取 20.00 mL 蒸馏水代替水样进行上述实验，测定空白值。

数据记录与处理

(1) 计算重铬酸钾标准溶液的浓度。

(2) 计算硫酸亚铁铵标准溶液的浓度、平均值和相对平均偏差($\bar{d}_r \leqslant 0.2\%$)。

$$c = \frac{10.00 \times 6 \times c_{K_2Cr_2O_7}}{V}$$

式中，c 为硫酸亚铁铵标准溶液的浓度，mol/L；V 为硫酸亚铁铵标准溶液的用量，mL。

(3) 计算水样的化学需氧量(COD_{Cr})。

$$COD_{Cr}(\text{以 } O_2 \text{ 计，mg/L}) = \frac{(V_1 - V_0) \times c \times 8 \times 1000}{V}$$

式中，c 为硫酸亚铁铵标准溶液的浓度，mol/L；V_0 为滴定水样时硫酸亚铁铵标准溶液的用量，mL；V_1 为滴定空白样时硫酸亚铁铵标准溶液的用量，mL；V 为水样的体积，mL；8 为 1/2 氧的摩尔质量，g/mol。

表 3.30　硫酸亚铁铵标准溶液的标定

	1	2	3
$V_{Fe(NH_4)_2(SO_4)_2}$(终)/mL			
$V_{Fe(NH_4)_2(SO_4)_2}$(始)/mL			

续表

	1	2	3
$V_{Fe(NH_4)_2(SO_4)_2}$ /mL			
$c_{Fe(NH_4)_2(SO_4)_2}$ / (mol/L)			
$\bar{c}_{Fe(NH_4)_2(SO_4)_2}$ / (mol/L)			
$\lvert d_i \rvert$			
$\bar{d}_r$ /%			

表 3.31　水样中 COD 的测定

	1	2	3
V_0/mL			
V_1/mL			
(V_1-V_0)/mL			
COD_{Cr} / (mg/L)			
$\overline{COD}_{Cr}$ / (mg/L)			
$\lvert d_i \rvert$			
$\bar{d}_r$ /%			

注意事项

(1) 水样取用体积可为 10.00～50.00 mL，试剂用量及浓度需按表 3.32 进行相应调整。

表 3.32　水样取用体积和试剂用量

水样体积/mL	0.25 mol/L 重铬酸钾/mL	硫酸-硫酸银/mL	硫酸亚铁铵/(mol/L)	硫酸汞/g	滴定前体积/mL
10.0	5.0	15	0.050	0.2	70
20.0	10.0	30	0.100	0.4	140
30.0	15.0	45	0.150	0.6	210
40.0	20.0	60	0.200	0.8	280
50.0	25.0	75	0.250	1.0	350

(2) 对于 COD 小于 50 mg/L 的水样，应改用 0.0250 mol/L 重铬酸钾标准溶液，回滴时用 0.01 mol/L 硫酸亚铁铵标准溶液。

(3) 水样加热回流后，溶液中重铬酸钾剩余量应为加入量的 1/5～4/5 为宜。

(4) 每次实验时应对硫酸亚铁铵标准溶液进行标定，室温较高时尤其应注意浓度的变化。

思考题

(1) 为什么需要做空白实验？

(2) 测定 COD_{Cr} 时有哪些干扰？如何消除这些干扰？

(3) 实验中，使用硫酸亚铁标准溶液应注意什么？

实验22 间接碘量法测定铜盐中的铜含量

实验目的

(1)掌握间接碘量法测定铜的原理。

(2)了解间接碘量法中误差的来源。

实验原理

在弱酸性(pH 为 3～4)条件下，Cu^{2+}与过量的 KI 作用，生成 CuI 沉淀和 I_2，析出的 I_2 以淀粉为指示剂，用 $Na_2S_2O_3$ 标准溶液滴定。反应式为

$$2Cu^{2+} + 4I^- \xlongequal{} 2CuI\downarrow + I_2$$

或

$$2Cu^{2+} + 5I^- \xlongequal{} 2CuI\downarrow + I_3^-$$

$$2S_2O_3^{2-} + I_3^- \xlongequal{} S_4O_6^{2-} + 3I^-$$

在以上反应中，I^-既是 Cu^{2+}的还原剂，又是 Cu^+的沉淀剂和 I_2 的络合剂。

间接碘量法须在弱酸性或中性溶液中进行。测定 Cu^{2+}时，通常用 NH_4HF_2 缓冲溶液(HF/F^- 共轭酸碱对)控制溶液的酸度为 3～4。同时 NH_4HF_2 缓冲溶液中的 F^-可以作为掩蔽剂，使共存的 Fe^{3+}转化为 FeF_6^{3-}，以消除其干扰。如试样中不含 Fe^{3+}，也可用乙酸缓冲溶液(pH ≈ 4)控制酸度。

CuI 沉淀表面易吸附少量 I_2，这部分 I_2 不与淀粉作用，会引起终点提前。因此，KSCN 溶液应在临近终点时加入，使 CuI 沉淀转化为溶解度更小的 CuSCN 沉淀。CuSCN 不吸附 I_2，从而使被 CuI 吸附的那部分 I_2 释放出来，提高测定的准确度。

实验用品

仪器：酸式滴定管(50 mL)，容量瓶(250 mL)，锥形瓶(250 mL)，碘量瓶(250 mL)，分析天平。

试剂：$Na_2S_2O_3$ 标准溶液(0.1 mol/L)，$K_2Cr_2O_7$ 标准溶液(0.017 mol/L)，KI 溶液(100 g/L，使用前配制)，KSCN 溶液(100 g/L)，H_2SO_4 溶液(1 mol/L)，HCl 溶液(6 mol/L)，淀粉溶液(5 g/L)，$CuSO_4 \cdot 5H_2O$ 试样。

实验步骤

(1) $Na_2S_2O_3$ 标准溶液(0.1 mol/L)的配制和标定。

称取 13 g $Na_2S_2O_3$ 溶于 250 mL 新煮沸并冷却的蒸馏水，溶解后，加入约 0.1 g Na_2CO_3，用新煮沸并冷却的蒸馏水稀释至 500 mL，储存于棕色试剂瓶中，在暗处放置 3～5 天后标定。

准确移取 25.00 mL $K_2Cr_2O_7$ 标准溶液(0.017 mol/L)于锥形瓶中，加入 5 mL HCl 溶液(6 mol/L)、5 mL KI 溶液(100 g/L)，摇匀，在暗处放置 5 min 让其反应完全，加入 50 mL 蒸馏水，用待标定的 $Na_2S_2O_3$ 标准溶液滴定至淡黄色，然后加入 3 mL 淀粉溶液(5 g/L)，继续滴定至溶液呈现亮绿色即为终点。将消耗 $Na_2S_2O_3$ 标准溶液的体积记录在表 3.33 中。平行滴定 3 次。

（2）铜盐中铜含量的测定。

准确称取 0.5～0.6 g $CuSO_4·5H_2O$ 试样，置于 250 mL 锥形瓶中，加 5 mL 1 mol/L H_2SO_4 溶液和 100 mL 水使其溶解。加入 10 mL KI 溶液（100 g/L），立即用 $Na_2S_2O_3$ 标准溶液滴定至溶液呈浅黄色。加入 2 mL 5 g/L 淀粉溶液，继续滴定至溶液呈浅蓝色。再加入 10 mL KSCN 溶液（100 g/L），溶液转为深蓝色。再继续用 $Na_2S_2O_3$ 标准溶液滴定至溶液蓝色刚好消失即为终点。将消耗 $Na_2S_2O_3$ 标准溶液的体积记录在表 3.34 中。平行滴定 3 次。

数据记录与处理

计算 $Na_2S_2O_3$ 标准溶液的浓度，计算试样中 Cu 的质量分数、平均值和相对平均偏差（$\bar{d}_r \leqslant 0.2\%$）。

表 3.33　$Na_2S_2O_3$ 标准溶液的标定

	1	2	3
$V_{Na_2S_2O_3}$（终）/mL			
$V_{Na_2S_2O_3}$（始）/mL			
$V_{Na_2S_2O_3}$ / mL			
$c_{Na_2S_2O_3}$ /（mol/L）			
$\bar{c}_{Na_2S_2O_3}$ /（mol/L）			
$\lvert d_i \rvert$			
$\bar{d}_r$ /%			

表 3.34　Cu 含量的测定

	1	2	3
$V_{Na_2S_2O_3}$（终）/mL			
$V_{Na_2S_2O_3}$（始）/mL			
$V_{Na_2S_2O_3}$ /mL			
c_{Cu}/（mol/L）			
w_{Cu}/%			
$\bar{w}_{Cu}$ /%			
$\lvert d_i \rvert$			
$\bar{d}_r$ /%			

注意事项

（1）在滴定过程中不断摇动样品溶液。

（2）溶液的 pH 一般控制在 3～4。酸度过低，Cu^{2+}容易水解；酸度太高，I^-易被空气中的氧气氧化。

思考题

（1）实验中加入 KI 的作用是什么？

（2）为什么要加入 KSCN？为什么不能过早地加入？

（3）若试样中含有铁，则滴定时加入何种试剂可以消除铁的干扰，并同时控制 pH 为 3～4？

实验 23 钙制剂中钙含量的测定

实验目的

(1) 掌握用高锰酸钾法测定钙的原理和方法。

(2) 了解沉淀分离的基本要求与操作。

实验原理

某些金属离子(如 Pb^{2+}、Cd^{2+}等)能与草酸根形成难溶的草酸盐沉淀。沉淀经过滤、洗净后，再用稀硫酸溶液溶解，然后用 $KMnO_4$ 标准溶液滴定释放出来的 $H_2C_2O_4$，即可间接测定这些金属离子的含量。以 Ca^{2+}为例，有关反应如下：

$$Ca^{2+} + C_2O_4^{2-} = CaC_2O_4\downarrow$$

$$CaC_2O_4 + 2H^+ = H_2C_2O_4 + Ca^{2+}$$

$$5H_2C_2O_4 + 2MnO_4^- + 6H^+ = 2Mn^{2+} + 10CO_2\uparrow + 8H_2O$$

用该法可测定某些钙制剂(如葡萄糖酸钙、钙片等)中的钙含量。

实验用品

仪器：酸式滴定管(50 mL)，移液管(25 mL)，容量瓶(250 mL)，锥形瓶(250 mL)，烧杯(100 mL)，分析天平，电热板。

试剂：$KMnO_4$ 标准溶液(0.02 mol/L)，$(NH_4)_2C_2O_4$ 溶液(0.05 mol/L)，$NH_3 \cdot H_2O$(7 mol/L)，HCl 溶液(6 mol/L)，H_2SO_4 溶液(1 mol/L)，甲基橙水溶液(1 g/L)，$AgNO_3$ 溶液(0.1 mol/L)，钙制剂。

实验步骤

称取约 0.05 g 钙制剂于 100 mL 烧杯中，加入适量蒸馏水及 2～5 mL HCl 溶液(6 mol/L)，轻轻摇动烧杯，加热溶解。稍冷后向溶液中加入 2～3 滴甲基橙(1 g/L)，再滴加氨水(7 mol/L)至溶液由红色变为黄色。趁热逐滴加入约 50 mL $(NH_4)_2C_2O_4$ 溶液(0.05 mol/L)，在低温电热板(或水浴)上陈化 30 min。冷却后过滤，将烧杯中的沉淀洗涤数次后转入漏斗中，继续洗涤沉淀至无 Cl^-(盛接洗涤液在 HNO_3 介质中以 $AgNO_3$ 检验)。将带有沉淀的滤纸铺在原烧杯的内壁上，用 50 mL H_2SO_4 溶液(1 mol/L)将沉淀由滤纸上洗入烧杯中，再用蒸馏水洗两次。加入蒸馏水稀释至 100 mL，加热至 70～80℃。用 $KMnO_4$ 标准溶液(0.02 mol/L，详见实验 18)滴定至溶液呈淡红色，再将滤纸搅入溶液中，若溶液褪色，则继续滴定，直至出现的淡红色 30 s 内不褪色即为终点。将消耗的 $KMnO_4$ 标准溶液的体积记录在表 3.35 中。平行滴定 3 份样品。

数据记录与处理

根据记录的数据计算钙制剂中钙的质量分数。

表 3.35　钙制剂中钙含量的测定

	1	2	3
$m_{试样}$/g			
V_{KMnO_4}(终)/mL			
V_{KMnO_4}(始)/mL			
V_{KMnO_4}/mL			
w_{Ca}/%			
$\overline{w}_{Ca}$/%			
$\|d_i\|$			
$\overline{d}_r$/%			

思考题

(1) 加入$(NH_4)_2C_2O_4$时，为什么要在热溶液中逐滴加入？

(2) 洗涤 CaC_2O_4 沉淀时，为什么要洗至无 Cl^-？

(3) 试比较 $KMnO_4$ 法和络合滴定法测定 Ca^{2+} 的优缺点。

实验 24　水中溶解氧的测定

实验目的

(1) 了解水中溶解氧的含义及测定的意义。

(2) 掌握碘量法测定水样中溶解氧的原理和方法。

实验原理

好氧性物质增多，使得水体中溶氧量降低，会造成大量鱼类窒息死亡，因而溶解氧(DO)的量可反映水体的受污染程度，也是评价水质的重要指标之一。本实验采用碘量法对溶解氧进行测定。在水样中加入硫酸锰及碱性碘化钾溶液，生成氢氧化锰沉淀。氢氧化锰极不稳定，迅速与水中溶解氧化合生成高价锰化合物。反应式如下：

$$MnSO_4 + 2NaOH = Mn(OH)_2\downarrow + Na_2SO_4$$

$$2Mn(OH)_2 + O_2 = 2H_2MnO_3$$

$$H_2MnO_3 + Mn(OH)_2 = Mn_2O_3\downarrow + 2H_2O$$

酸性条件下，高价锰氧化物将碘化钾氧化并释放出与溶解氧量相当的游离碘。然后用硫代硫酸钠标准溶液滴定，换算出溶解氧的含量。反应式如下：

$$2KI + H_2SO_4 = 2HI + K_2SO_4$$

$$Mn_2O_3 + 2H_2SO_4 + 2HI = 2MnSO_4 + I_2 + 3H_2O$$

$$I_2 + 2Na_2S_2O_3 = 2NaI + Na_2S_4O_6$$

实验用品

仪器：碱式滴定管(50 mL)，移液管(25 mL)，容量瓶(1000 mL)，碘量瓶(250 mL)，锥形瓶(250 mL)。

试剂：H_2SO_4溶液(1∶5，在烧杯中加入 5 体积蒸馏水，在搅拌下加入 1 体积浓硫酸)，硫酸锰溶液(称取 480 g $MnSO_4 \cdot 4H_2O$ 或 364 g $MnSO_4 \cdot H_2O$ 溶于蒸馏水中，过滤并稀释至 1000 mL)，碱性碘化钾溶液[称取 500 g NaOH 溶于 300～400 mL 蒸馏水中，另称取 150 g KI (或 135 g NaI)溶于 200 mL 蒸馏水中，待 NaOH 溶液冷却后，将两溶液合并混匀，用蒸馏水稀释至 1000 mL。静置 24 h，倒出上层澄清液，储于棕色瓶中。用橡皮塞塞紧，避光保存]，淀粉溶液(1%，称取 1 g 可溶性淀粉，用少量水调成糊状，用刚煮沸的水稀释至 100 mL)，重铬酸钾标准溶液(0.025 mol/L)，硫代硫酸钠溶液。

实验步骤

(1) 重铬酸钾标准溶液(0.025 mol/L)的配制。

称取 1.84 g $K_2Cr_2O_7$(105～110℃烘干 2 h)，溶于蒸馏水中，转移至 250 mL 容量瓶中，用水稀释至刻度，摇匀。

(2) 硫代硫酸钠标准溶液(0.025 mol/L)的配制和标定。

称取 6.2 g 硫代硫酸钠($Na_2S_2O_3 \cdot 5H_2O$)，溶于 1000 mL 蒸馏水中，加入 0.4 g NaOH 或 0.2 g Na_2CO_3，储于棕色瓶中。此溶液浓度约为 0.025 mol/L。

准确浓度可按下法标定：于 250 mL 碘量瓶中加入 100 mL 蒸馏水和 1 g KI，用移液管吸取 10.00 mL $K_2Cr_2O_7$标准溶液(0.025 mol/L)、5 mL H_2SO_4溶液(1∶5)，加塞摇匀。置于暗处 5 min，取出后用待标定的硫代硫酸钠标准溶液滴定至溶液由棕色变为淡黄色时，加入 1 mL 淀粉溶液，继续滴定至蓝色刚好褪色为止。在表 3.36 中记录消耗 $Na_2S_2O_3$标准溶液的体积。

(3) 水样的采集。

先用水样冲洗溶解氧瓶后，沿瓶壁直接注入水样或用虹吸法将细管插入溶解氧瓶底部，注入水样至溢流出瓶容积的 1/3～1/2。不要使水样曝气或有气泡残存在溶解氧瓶中。

(4) 溶解氧的固定。

移取 1.00 mL $MnSO_4$溶液，加入装有水样的溶解氧瓶中，加注时，应将吸量管插入液面下。按上法，加入 2 mL 碱性 KI 溶液。盖紧瓶塞，将瓶颠倒混合数次，静置。待沉淀降至瓶内一半时，再颠倒混合一次，待沉淀物下降至瓶底。一般在取样现场固定溶解氧。

(5) 碘的析出。

轻轻打开瓶塞，立即用吸量管吸取 2.00 mL 浓 H_2SO_4，插入液面下加入，盖紧瓶塞。颠倒混合，直至沉淀物全部溶解为止。放置暗处 5 min。

(6) 样品的测定。

移取 100.00 mL 上述溶液于 250 mL 锥形瓶中，用 $Na_2S_2O_3$标准溶液(0.025 mol/L)滴定至溶液呈淡黄色。加入 1 mL 淀粉溶液，继续滴定至蓝色刚刚褪去。在表 3.37 中记录消耗硫代硫酸钠标准溶液的体积。平行滴定 3 次。

数据记录与处理

(1) 计算硫代硫酸钠标准溶液的浓度、平均值和相对平均偏差($\bar{d}_r \leqslant 0.2\%$)。

$$c = \frac{10.00 \times 6 \times m}{V \times M \times 0.25}$$

式中：V 为消耗硫代硫酸钠标准溶液的体积，mL；m 为称取的重铬酸钾的质量，g；M 为重铬酸钾的摩尔质量，g/mol。

(2) 计算溶解氧的含量。

$$c_{O_2} = \frac{c \times V \times 8 \times 1000}{100}$$

式中：c_{O_2} 为水样中溶解氧的浓度，mg/L；c 为硫代硫酸钠标准溶液的浓度，mol/L；V 为消耗硫代硫酸钠标准溶液的体积，mL；8 为 1/2 氧的摩尔质量，g/mol；100 为水样的体积，mL。

表 3.36　$Na_2S_2O_3$ 标准溶液的标定

	1	2	3
$V_{Na_2S_2O_3}$ (终)/mL			
$V_{Na_2S_2O_3}$ (始)/mL			
$V_{Na_2S_2O_3}$ /mL			
$c_{Na_2S_2O_3}$/ (mol/L)			
$\bar{c}_{Na_2S_2O_3}$ /(mol/L)			
$\lvert d_i \rvert$			
$\bar{d}_r$ /%			

表 3.37　溶解氧的测定

	1	2	3
$V_{Na_2S_2O_3}$ (终)/mL			
$V_{Na_2S_2O_3}$ (始)/mL			
$V_{Na_2S_2O_3}$ /mL			
c_{O_2} /(mg/L)			
$\bar{c}_{O_2}$ /(mg/L)			

注意事项

(1) 当水样中含有亚硝酸盐时会干扰测定，可将叠氮化钠加入碱性碘化钾溶液中使亚硝酸盐分解而消除干扰。

(2) 如水样中含氧化性物质(如游离氯等)，应预先加入相当量的硫代硫酸钠除去。

(3) 当水样中三价铁离子的浓度达 100～200 mg/L 时，可加入 1 mL 40%氟化钾溶液消除干扰。

思考题

(1) 为什么要先滴定至淡黄色再加淀粉溶液？可以在溶液深黄色时就加吗？

(2) 用碘量法测定溶解氧受哪些因素影响？如样品中含有氧化物、藻类、悬浮物将产生什么干扰？

实验 25　果汁中维生素 C 含量的测定

实验目的

(1) 掌握碘标准溶液的配制和标定。

(2) 掌握碘量法测定维生素 C 的原理和方法。

实验原理

维生素 C (Vc) 又称抗坏血酸，分子式为 $C_6H_8O_6$，维生素 C 中的烯二醇具有还原性，能被 I_2 定量氧化成二酮基。因此，可以淀粉溶液为指示剂，用 I_2 溶液滴定维生素 C。反应式如下：

$$C_6H_8O_6 + I_2 \!=\!=\!= C_6H_6O_6 + 2HI$$

由于维生素 C 的还原性很强，在空气中容易被氧化，尤其是在碱性介质中，因此测定时加入乙酸使溶液呈弱酸性，以降低氧化速度。

实验用品

仪器：碱式滴定管 (50 mL)，移液管 (25 mL)，锥形瓶 (250 mL)，研钵，棕色试剂瓶 (250 mL)。

试剂：$Na_2S_2O_3 \cdot 5H_2O$ (s)，I_2 标准溶液 (0.05 mol/L)，淀粉溶液 (0.5%，称取 0.5 g 可溶性淀粉，用少量水调成糊状，慢慢加入 100 mL 微沸的蒸馏水中，继续煮沸至溶液透明为止)，HAc 溶液 (2 mol/L)，果汁。

实验步骤

(1) $Na_2S_2O_3$ 标准溶液 (0.05 mol/L) 的配制和标定。

详见实验 22。

(2) I_2 标准溶液 (0.05 mol/L) 的配制和标定。

称取 3.3 g I_2 和 5 g KI 置于研钵中，加少量水研磨，待 I_2 全部溶解后，将溶液转入棕色瓶中，加水稀释至 250 mL，充分摇匀，放暗处保存。准确移取 25.00 mL 已标定的 $Na_2S_2O_3$ 标准溶液于 250 mL 锥形瓶中，加 50 mL 蒸馏水、2 mL 淀粉溶液，用 I_2 标准溶液滴定至稳定的浅蓝色，30 s 内不褪色即为终点。将消耗 I_2 标准溶液的体积记录在表 3.38 中。平行滴定 3 次。

(3) 果汁中维生素 C 含量的测定。

准确移取 25.00 mL 果汁于 250 mL 锥形瓶中，加入 10 mL HAc 溶液 (2 mol/L)、2 mL 淀粉溶液，混匀。立即用 I_2 标准溶液滴定至稳定的浅蓝色。将消耗 I_2 标准溶液的体积记录在表 3.39 中。平行滴定 3 次。

数据记录与处理

(1) 计算 I_2 标准溶液的浓度、平均值和相对平均偏差。

(2) 计算果汁中维生素 C 的含量。

$$\rho_{Vc}=\frac{cV_1}{V}\times 176.12\times 1000$$

式中，ρ_{Vc} 为果汁中维生素 C 的浓度，g/L；c 为 I_2 标准溶液的浓度，mol/L；V_1 为消耗 I_2 标准溶液的体积，mL；V 为果汁的体积，mL；176.12 为维生素 C 的摩尔质量，g/mol。

表 3.38　I_2 标准溶液的标定

	1	2	3
V_{I_2} (终)/mL			
V_{I_2} (始)/mL			
V_{I_2} /mL			
c_{I_2}/ (mol/L)			
$\bar{c}_{I_2}$ / (mol/L)			
$\lvert d_i \rvert$			
$\bar{d}_r$ /%			

表 3.39　维生素 C 含量的测定

	1	2	3
V_{I_2} (终)/mL			
V_{I_2} (始)/mL			
V_{I_2} /mL			
ρ_{Vc}/ (g/L)			
$\bar{\rho}_{Vc}$ / (g/L)			
$\lvert d_i \rvert$			
$\bar{d}_r$ /%			

注意事项

(1) 维生素 C 很容易被氧化，整个实验过程动作要快。

(2) 有些较黏稠的果汁可稀释、过滤后再测定。

思考题

(1) 样品中加入乙酸的作用是什么？

(2) 配制碘溶液时加入碘化钾的目的是什么？

实验 26　葡萄糖注射液中葡萄糖含量的测定

实验目的

了解碘量法测定葡萄糖的原理和方法。

实验原理

在碱性溶液中，I_2 可歧化成 IO^-和 I^-，IO^-能定量地将葡萄糖($C_6H_{12}O_6$)氧化成葡萄糖酸($C_6H_{12}O_7$)，未与 $C_6H_{12}O_6$ 反应的 IO^-进一步歧化为 IO_3^-和 I^-。溶液酸化后，IO_3^-又与 I^-反应析出 I_2，用 $Na_2S_2O_3$ 标准溶液滴定析出的 I_2，由此可计算出 $C_6H_{12}O_6$ 的含量。有关反应式如下：

$$I_2 + 2OH^- = IO^- + I^- + H_2O$$

$$C_6H_{12}O_6 + IO^- = I^- + C_6H_{12}O_7$$

总反应式为

$$I_2 + C_6H_{12}O_6 + 2OH^- = C_6H_{12}O_7 + 2I^- + H_2O$$

剩下未反应的 IO^-在碱性条件下发生歧化反应：

$$3IO^- = IO_3^- + 2I^-$$

在酸性条件下

$$IO_3^- + 5I^- + 6H^+ = 3I_2 + 3H_2O$$

即

$$IO^- + I^- + 2H^+ = I_2 + H_2O$$

$$I_2 + 2S_2O_3^{2-} = 2I^- + S_4O_6^{2-}$$

由以上反应可以看出，一分子葡萄糖与一分子 I_2 相当。

实验用品

仪器：酸式滴定管(50 mL)，容量瓶(250 mL)，锥形瓶(250 mL)，分析天平。

试剂：HCl 溶液(2 mol/L)，NaOH 溶液(0.2 mol/L)，$Na_2S_2O_3$ 标准溶液(0.05 mol/L)，I_2 标准溶液(0.05 mol/L)，淀粉溶液(0.5%)，KI(s)，葡萄糖注射液(0.5%，将 5%的葡萄糖注射液稀释 10 倍)。

实验步骤

(1) $Na_2S_2O_3$ 标准溶液(0.05 mol/L)的配制和标定。

详见实验 22。

(2) I_2 标准溶液(0.05 mol/L)的配制和标定。

详见实验 25。

(3) 葡萄糖含量的测定。

移取 25.00 mL 葡萄糖注射液于 250 mL 容量瓶中，加水稀释至刻度，摇匀。移取 25.00 mL 稀释后的葡萄糖溶液于 250 mL 锥形瓶中，准确加入 25.00 mL I_2 标准溶液，慢慢滴加

0.2 mol/L NaOH，边加边摇，直至溶液呈淡黄色(加碱的速度不能过快，否则生成的IO^-来不及氧化 $C_6H_{12}O_6$，使测定结果偏低)。用表面皿将锥形瓶盖好，放置 10～15 min，然后加 6 mL HCl(2 mol/L)使溶液呈酸性，并立即用 $Na_2S_2O_3$ 标准溶液滴定，至溶液呈浅黄色时，加入 3 mL 淀粉溶液，继续滴定至蓝色刚好消失。平行滴定 3 次。在表 3.40 中记录消耗 $Na_2S_2O_3$ 标准溶液的体积。

数据记录与处理

根据记录的实验数据计算样品中葡萄糖的含量。

表 3.40　葡萄糖注射液中葡萄糖含量的测定

	1	2	3
$V_{Na_2S_2O_3}$ (终)/mL			
$V_{Na_2S_2O_3}$ (始)/mL			
$V_{Na_2S_2O_3}$ /mL			
ρ /(g/mL)			
$\bar{\rho}$ /(g/mL)			
$\|d_i\|$			
$\bar{d}_r$ /%			

思考题

(1)配制 I_2 溶液时为什么要加入 KI？
(2)碘量法的主要误差来源有哪些？如何消除误差？

3.5　沉淀滴定及重量分析法实验

实验 27　莫尔法测定可溶性氯化物中的氯含量

实验目的

(1)掌握 $AgNO_3$ 标准溶液的配制和标定方法。
(2)掌握莫尔法的原理和方法。

实验原理

在中性或弱碱性溶液中，以 K_2CrO_4 为指示剂，用 $AgNO_3$ 标准溶液滴定试液中的 Cl^-。由于 AgCl 的溶解度小于 Ag_2CrO_4，溶液中首先析出 AgCl 沉淀，计量点后稍过量的 Ag^+与 CrO_4^{2-}反应生成砖红色 Ag_2CrO_4 沉淀而指示终点。主要反应如下：

$$Ag^+ + Cl^- = AgCl\downarrow \text{(白色)} \qquad K_{sp}=1.8\times10^{-10}$$

$$2Ag^+ + CrO_4^{2-} = Ag_2CrO_4\downarrow \text{(砖红色)} \qquad K_{sp}=2.0\times10^{-12}$$

根据消耗 $AgNO_3$ 标准溶液的体积和浓度，可计算试样中氯的含量。

实验用品

仪器：酸式滴定管(50 mL)，移液管(25 mL)，容量瓶(100 mL，250 mL)，锥形瓶(250 mL)，分析天平，棕色试剂瓶(500 mL)。

试剂：$AgNO_3$(AR)，NaCl(GR，500～600℃下干燥 2～3 h，储存于干燥器中)，K_2CrO_4 溶液(50 g/L)，NaCl 试样。

实验步骤

(1) $AgNO_3$ 标准溶液(0.1 mol/L)的配制和标定。

称取 8.5 g $AgNO_3$ 于小烧杯中，用蒸馏水溶解后，转移至棕色试剂瓶中，稀释至 500 mL，盖上瓶塞，置于暗处保存。

准确称取 0.55～0.60 g NaCl 基准物质于小烧杯中，用蒸馏水溶解后，转移至 100 mL 容量瓶中，用水稀释至刻度，摇匀。准确移取 25.00 mL NaCl 溶液于 250 mL 锥形瓶中，加入 20 mL 蒸馏水、1 mL K_2CrO_4 溶液(50 g/L)。用 $AgNO_3$ 溶液滴定至溶液呈橙红色即为终点。将消耗 $AgNO_3$ 标准溶液的体积记录于表 3.41 中。平行滴定 3 次。

(2) 试样中氯含量的测定。

称取 1.6 g NaCl 试样于小烧杯中，用蒸馏水溶解后，转移至 250 mL 容量瓶中，用水稀释至刻度，摇匀。准确移取 25.00 mL 试液于 250 mL 锥形瓶中，加 20 mL 蒸馏水、1 mL K_2CrO_4 溶液(50 g/L)。用 $AgNO_3$ 标准溶液滴定至溶液呈橙红色即为终点。将消耗 $AgNO_3$ 标准溶液的体积记录于表 3.42 中。平行滴定 3 次。

数据记录与处理

(1) 计算 $AgNO_3$ 标准溶液的浓度、平均值和相对平均偏差($\bar{d}_r \leqslant 0.2\%$)。

$$c_{AgNO_3} = \frac{m_{NaCl} \times \frac{25.00}{100.0}}{M_{NaCl} V_{AgNO_3}} \times 1000$$

表 3.41　$AgNO_3$ 标准溶液的标定

	1	2	3
V_{AgNO_3} (终)/mL			
V_{AgNO_3} (始)/mL			
V_{AgNO_3} /mL			
c_{AgNO_3}/ (mol/L)			
$\bar{c}_{AgNO_3}$/(mol/L)			
$\lvert d_i \rvert$			
$\bar{d}_r$ /%			

表 3.42　试样中氯含量的测定

	1	2	3
m_s/g			
V_{AgNO_3}（终）/mL			
V_{AgNO_3}（始）/mL			
V_{AgNO_3} /mL			
w_{Cl}/%			
$\bar{w}_{Cl}$/%			
$\lvert d_i \rvert$			
$\bar{d}_r$ /%			

(2) 计算试样中氯的质量分数。

$$w_{Cl}=\frac{c_{AgNO_3}V_{AgNO_3}}{m_s\times\frac{25.00}{250.0}}\times\frac{M_{Cl}}{1000}\times 100\% \qquad (M_{NaCl}=58.44\ g/mol,\ M_{AgNO_3}=169.88\ g/mol)$$

注意事项

(1) 滴定时应剧烈摇动，使 AgCl 沉淀吸附的 Cl^-及时释放出来，防止终点提前。

(2) 滴定必须在中性或弱碱性溶液中进行，最佳 pH 范围为 6.5～10.5。若有铵盐存在，溶液 pH 范围应控制在 6.5～7.2 为宜。

(3) 指示剂的用量对实验结果有影响，一般以 5×10^{-3} mol/L 为宜。

(4) 实验结束后，盛装 $AgNO_3$ 溶液的滴定管应先用蒸馏水冲洗 2～3 次，再用自来水冲洗，以免产生 AgCl 沉淀。含银废液应予以回收。

思考题

(1) 以 K_2CrO_4 作指示剂时，其浓度太大或太小对测定有何影响？

(2) 配制好的 $AgNO_3$ 溶液为什么要储存于棕色瓶中，并置于暗处？

(3) 能否用莫尔法以 NaCl 标准溶液直接滴定 Ag^+？为什么？

实验 28　福尔哈德法测定可溶性氯化物中的氯含量

实验目的

(1) 掌握 NH_4SCN 标准溶液的配制和标定方法。

(2) 掌握用福尔哈德法测定氯化物中氯含量的原理和方法。

实验原理

在含 Cl^-的酸性溶液中加入过量 $AgNO_3$ 标准溶液，定量生成 AgCl 沉淀。然后以铁铵矾作指示剂，用 NH_4SCN 标准溶液返滴定过量的 Ag^+，由$[Fe(SCN)]^{2+}$的红色指示滴定终点(当指示剂用量少时呈橙色)。反应式如下：

$$Ag^{+}+Cl^{-} = AgCl\downarrow \text{（白色）} \qquad K_{sp}=1.8\times10^{-10}$$

$$Ag^{+}+SCN^{-} = AgSCN\downarrow \text{（白色）} \qquad K_{sp}=1.0\times10^{-12}$$

$$Fe^{3+}+SCN^{-} = [Fe(SCN)]^{2+} \text{（红色）} \qquad K_{1}=138$$

设 $AgNO_3$ 标准溶液的浓度为 c(mol/L)，标定 NH_4SCN 标准溶液时消耗 NH_4SCN 标准溶液的体积为 V(mL)，测定试样中氯含量时加入 $AgNO_3$ 标准溶液的体积为 V_1(mL)，返滴定时消耗 NH_4SCN 标准溶液的体积为 V_2(mL)，则试样中氯的质量分数可用下式计算:

$$c_{NH_4SCN}=\frac{25.00\times c_{AgNO_3}}{V}$$

$$w_{Cl}=\frac{35.45\times(c_{AgNO_3}\times V_1-c_{NH_4SCN}\times V_2)\times250}{1000\times m_s\times25.00}\times100\%$$

实验用品

仪器：酸式滴定管(50 mL)，容量瓶(250 mL，500 mL)，移液管(25 mL)，锥形瓶(250 mL)，烧杯(100 mL)，量筒(5 mL，25 mL)，分析天平，棕色试剂瓶。

试剂：$AgNO_3$ 标准溶液(0.10 mol/L)，NH_4SCN 标准溶液(0.10 mol/L)，铁铵矾指示剂(400 g/L)，HNO_3 溶液(8 mol/L)，硝基苯，NaCl 试样。

实验步骤

(1) $AgNO_3$ 标准溶液(0.10 mol/L)的配制和标定。

详见实验 27。

(2) NH_4SCN 标准溶液(0.10 mol/L)的配制和标定。

称取 3.8 g NH_4SCN 于 100 mL 烧杯中，加水溶解，转移至 500 mL 容量瓶中，定容，摇匀后转入试剂瓶中储存。

移取 25.00 mL $AgNO_3$ 标准溶液于 250 mL 锥形瓶中，加入 5 mL HNO_3 溶液及 1.0 mL 铁铵矾指示剂，然后用 NH_4SCN 标准溶液滴定。滴定时，充分振荡溶液，滴定至溶液呈稳定的淡红色即为终点。在表 3.43 中记录消耗 NH_4SCN 标准溶液的体积 V(mL)。平行滴定 3 次。

(3) 试样中氯含量的测定。

准确称取 1.6 g NaCl 试样于 100 mL 烧杯中，用蒸馏水溶解，转移至 250 mL 容量瓶中，用蒸馏水稀释至刻度，摇匀。准确移取 25.00 mL 试样溶液于 250 mL 锥形瓶中，加 20 mL 蒸馏水、5 mL HNO_3 溶液。用滴定管加入 $AgNO_3$ 标准溶液至过量 5～10 mL，准确记录加入的 $AgNO_3$ 标准溶液的体积 V_1(mL)。然后加入 2 mL 硝基苯，用橡皮塞塞住瓶口，剧烈振荡 30 s，使 AgCl 沉淀进入硝基苯层而与溶液分离。再加入 1.0 mL 铁铵矾指示剂，用 NH_4SCN 标准溶液滴定至稳定的淡红色即为终点。在表 3.44 中记录消耗 NH_4SCN 标准溶液的体积 V_2(mL)。平行滴定 3 次。

数据记录与处理

计算 NH_4SCN 标准溶液的浓度及试样中的氯含量。

表 3.43　NH_4SCN 标准溶液的标定

	1	2	3
m_{NH_4SCN} /g			
V(终) /mL			
V(始) /mL			
V/mL			
c_{NH_4SCN} /(mol/L)			
$\bar{c}_{NH_4SCN}$ /(mol/L)			
$\lvert d_i \rvert$			
$\bar{d}_r$ /%			

表 3.44　试样中氯含量的测定

	1	2	3
m_s /g			
V_1/mL			
V_2(终) /mL			
V_2(始) /mL			
V_2/mL			
w_{Cl}/%			
$\bar{w}_{Cl}$/%			
$\lvert d_i \rvert$			
$\bar{d}_r$ /%			

注意事项

(1) 指示剂用量对滴定有影响，一般控制 Fe^{3+}浓度为 0.015 mol/L 为宜。

(2) 滴定时应控制 H^+浓度为 0.1～1 mol/L，同时剧烈振荡溶液。

(3) 为了防止 AgCl 沉淀与 SCN^-发生反应消耗滴定剂，滴定前应加入硝基苯(有毒)或石油醚以保护 AgCl 沉淀，使其不与 SCN^-接触。

思考题

(1) 为什么用 HNO_3 酸化溶液？可否用 HCl 或 H_2SO_4 酸化？为什么？

(2) 试讨论酸度对福尔哈德法测定卤素离子含量的影响。

实验 29　重量分析法测定钡盐中的钡含量

实验目的

(1) 了解沉淀法的条件和方法。

(2) 掌握重量分析法的基本操作。

实验原理

Ba^{2+} 可生成一系列难溶化合物，如 $BaCO_3$、$BaCrO_4$、$BaSO_4$、BaC_2O_4 等。其中，$BaSO_4$ 的溶解度最小（100 mL 溶液中，100℃时溶解 0.4 mg，25℃时仅溶解 0.25 mg），组成与化学式符合，摩尔质量较大，性质稳定，符合重量分析对沉淀的要求。

为了获得颗粒较大且纯净的 $BaSO_4$ 晶形沉淀，试样溶于水后，加稀 HCl 溶液酸化，加热至微沸。在不断搅动的条件下，慢慢地加入稀、热的 H_2SO_4 溶液，Ba^{2+}与 SO_4^{2-}反应形成晶形沉淀。沉淀经陈化、过滤、洗涤、干燥、炭化、灼烧后，以 $BaSO_4$ 形式称量。

实验用品

仪器：烧杯（250 mL），量筒（100 mL），漏斗，定量滤纸，玻璃棒，坩埚，马弗炉，表面皿，分析天平。

试剂：$BaCl_2$ 试样，HCl 溶液（2 mol/L），H_2SO_4 溶液（1 mol/L），HNO_3 溶液（2 mol/L），$AgNO_3$ 溶液（0.1 mol/L）。

实验步骤

（1）样品处理。

称取 0.4～0.5 g $BaCl_2 \cdot 2H_2O$ 试样于 250 mL 烧杯中，加入 100 mL 蒸馏水、4 mL HCl 溶液（2 mol/L），搅拌溶解，加热至近沸。

（2）沉淀。

取 4 mL H_2SO_4 溶液（1 mol/L）于 100 mL 烧杯中，加 30 mL 蒸馏水，加热至近沸，趁热将 H_2SO_4 溶液用滴管逐滴加入 $BaCl_2$ 溶液中，并不断搅拌。

（3）陈化。

用 H_2SO_4 溶液（0.1 mol/L）检查上清液，仔细观察沉淀是否完全。沉淀完全后盖上表面皿，在沸腾的水浴上陈化半小时，期间搅动几次。

（4）过滤和洗涤。

将慢速定量滤纸按漏斗角度的大小折叠好，使其与漏斗很好地贴合。用水润湿，并使漏斗颈内保持水柱。将漏斗置于漏斗架上，漏斗下面放一个干净的烧杯。小心地将上层清液沿玻璃棒倾入漏斗中，再用倾泻法洗涤沉淀 3～4 次，每次用 15～20 mL 洗涤液（取 3 mL 1 mol/L H_2SO_4 溶液，加水稀释至 200 mL），最后将沉淀定量转移至滤纸上，洗涤沉淀至无 Cl^-为止（$AgNO_3$ 溶液检查）。

（5）炭化、灰化、灼烧。

将洁净带盖的坩埚在 800～850℃灼烧至恒量（m_1）。将折叠好的沉淀滤纸包置于已恒量的坩埚中，用马弗炉烘干、炭化后，于 800～850℃灼烧至恒量（m_2）。平行测定两份样品，将质量 m_1(g) 和 m_2(g) 记录在表 3.45 中。

数据记录与处理

计算试样中的钡含量。

$$w_{Ba}=\frac{\frac{M_{Ba}}{M_{BaSO_4}}\times(m_2-m_1)}{m_{BaCl_2\cdot 2H_2O}}\times 100\%$$

表 3.45 钡含量的测定

	1	2
$m_{试样}$/g		
m_1/g		
m_2/g		
(m_2-m_1)/g		
w_{Ba}/%		
$\bar{w}_{Ba}$/%		
$\|d_i\|$		
$\bar{d}_r$/%		

注意事项

(1)整个过程中，玻璃棒直至过滤、洗涤完毕后才能取出。

(2)在沉淀过程中，滴加速度不能过快，并需不断搅拌以免局部浓度过高，同时也可减少杂质的吸附现象。

(3)搅拌时，玻璃棒不要碰触杯壁及杯底，以免划伤烧杯，使沉淀黏附在烧杯划痕内，难以洗涤。

(4)陈化过程中，应将烧杯斜放在小木块上，使沉淀下沉并集中在烧杯一侧，以利于沉淀的分离和转移。

思考题

(1)为什么要在稀热 HCl 溶液中沉淀 $BaSO_4$？HCl 加入太多有什么影响？

(2)为什么要在热溶液中沉淀 $BaSO_4$，但要在冷却后过滤？沉淀为什么要陈化？

(3)用洗涤液洗涤 $BaSO_4$ 时，为什么要少量多次？

(4)为什么要准确称取 0.4～0.5 g $BaCl_2\cdot 2H_2O$ 试样？过多或过少有什么影响？

实验 30 钡盐中钡含量的测定(微波干燥重量法)

实验目的

学习利用微波炉干燥恒量 $BaSO_4$ 沉淀测定可溶性盐中的钡含量。

实验原理

实验原理及沉淀条件与实验 29 基本相同。用微波炉干燥恒量 $BaSO_4$ 沉淀时，试样内外同时加热，没有传热过程，加热迅速、均匀，瞬时可达较高温度。若沉淀中包藏有 H_2SO_4 等高沸点杂质，利用微波加热干燥 $BaSO_4$ 沉淀过程中杂质难以分解或挥发。因此，在沉淀条件和洗涤操作方面，要将含 Ba^{2+} 的试液进一步稀释，过量沉淀剂(H_2SO_4)控制在 20%～50%。

实验用品

仪器：烧杯(100 mL，250 mL)，玻璃棒，微波炉，循环水真空泵(配抽滤瓶)，G4 砂芯坩埚。

试剂：H_2SO_4 溶液(1 mol/L，0.1 mol/L)，HCl 溶液(2 mol/L)，HNO_3 溶液(2 mol/L)，$AgNO_3$ 溶液(0.1 mol/L)，$BaCl_2 \cdot 2H_2O$。

实验步骤

(1) 沉淀的制备。

详见实验 29。

(2) 沉淀的处理。

新制备的 $BaSO_4$ 沉淀陈化后，用已经在微波炉中恒量的 G4 砂芯坩埚(m_1)减压过滤并洗涤。然后将盛沉淀的坩埚置于微波炉内干燥(第一次 10 min，第二次 4 min)。干燥后，将坩埚转入干燥器中冷却至室温(10～15 min)，称量，重复操作直至恒量(m_2)。将质量 m_1(g) 和 m_2(g) 记录于表 3.46 中。

数据记录与处理

计算试样中的钡含量。

表 3.46　钡含量的测定

	1	2
$m_{试样}$/g		
m_1/g		
m_2/g		
(m_2-m_1)/g		
w_{Ba}/%		
$\bar{w}_{Ba}$/%		
$\vert d_i \vert$		
$\bar{d}_r$/%		

注意事项

(1) 洁净的 G4 砂芯坩埚使用前用真空泵抽 2 min 以除去玻璃砂板微孔中的水分，便于干燥。将其置于微波炉中，于 500 W(中高温)的输出功率下进行干燥，第一次 10 min，第二次 4 min。每次干燥后置于干燥器中冷却 10～15 min(刚放进时留一小缝隙，约 30 s 后再盖严)，然后在分析天平上快速称量。要求两次干燥后称量所得质量之差不超过 0.4 mg(已恒量)。

(2) 循环水真空泵及微波炉的使用方法与注意事项由指导教师讲授或参考有关说明书。

思考题

(1) 微波加热技术在分析化学(如分解试样和烘干试样等)中的应用有哪些优越性?

(2) 如何科学合理地进行本实验,以充分体现微波加热技术在重量分析中的应用特点?

实验 31 肥料中钾含量的测定

实验目的

(1) 了解肥料试样溶液的制备方法。

(2) 学习以四苯硼钠为沉淀剂测定钾含量的重量分析法。

实验原理

肥料试样经处理后,加入四苯硼钠试剂,产生四苯硼钾沉淀。反应式如下:

$$Na[B(C_6H_5)_4] + K^+ \xlongequal{} K[B(C_6H_5)_4] \downarrow + Na^+$$

所得 $K[B(C_6H_5)_4]$沉淀具有溶解度小、热稳定性较好等优点。沉淀生成后,经过一系列处理,称量并换算成 K_2O 的质量。四苯硼钾沉淀在碱性介质中进行,铵离子的干扰可用甲醛掩蔽,金属离子的干扰可用乙二胺四乙酸二钠掩蔽。

实验用品

仪器:烧杯(250 mL),容量瓶(100 mL),移液管(25 mL),量筒(5 mL, 10 mL),表面皿,G4 砂芯坩埚,分析天平。

试剂:甲醛溶液(25 g/L),乙二胺四乙酸二钠溶液(0.1 mol/L),酚酞指示剂(10 g/L),NaOH 溶液(20 g/L),四苯硼钾饱和溶液(过滤至清亮为止),四苯硼钠溶液[0.1 mol/L,称取 3.3 g 四苯硼钠,溶于 100 mL 蒸馏水中,加入 1 g $Al(OH)_3$,搅匀,放置过夜,反复过滤至清亮为止],浓 HCl,HNO_3 溶液(1 mol/L)。

实验步骤

(1) 肥料试样溶液的制备。

准确称取约 0.5 g 无机肥料于 250 mL 烧杯中,加入 20~30 mL 蒸馏水和 5~6 滴浓 HCl,盖上表面皿,低温煮沸 10 min。冷却后,将烧杯内残渣及溶液过滤于 100 mL 容量瓶中,用热蒸馏水洗涤烧杯内壁 5~6 次,滤液转入同一容量瓶中,用蒸馏水稀释至刻度,摇匀备用。

(2) 样品分析。

移取 10~25 mL 肥料制备液(根据试样中钾含量而定)于 250 mL 烧杯中,加入 5 mL 甲醛溶液(25 g/L)和 10 mL 乙二胺四乙酸二钠溶液(0.1 mol/L)。搅匀后加入 2 滴酚酞指示剂(10 g/L),用 NaOH 溶液(20 g/L)滴定至溶液呈淡红色。然后加热至 40℃,逐滴加入 5 mL 四苯硼钠溶液(0.1 mol/L),并搅拌 2~3 min,静置 30 min 后,用已恒量的 G4 砂芯坩埚(m_1)过滤，用四苯硼钾饱和溶液洗涤 2～3 次，最后用蒸馏水洗涤 3～4 次(每次约 5 mL)，抽滤至干。将坩埚置于干燥箱(或烘箱)中，120℃干燥 1 h 后放入干燥器中，冷却至室温后称量，再烘干、冷却、称量，直至恒量(m_2)。将质量 m_1(g)和 m_2(g)记录于表 3.47 中。

数据记录与处理

根据四苯硼钾沉淀的质量计算肥料中 K_2O 的质量分数。

表 3.47　K_2O 含量的测定

	1	2
$m_{试样}$/g		
m_1/g		
m_2/g		
(m_2-m_1)/g		
w_{K_2O}/%		
$\bar{w}_{K_2O}$/%		
$\|d_i\|$		
$\bar{d}_r$/%		

思考题

(1) 加入四苯硼钠溶液之前为什么加入 NaOH 溶液？

(2) 测定过程中为什么要加入甲醛和乙二胺四乙酸二钠溶液？

(3) 为什么要用四苯硼钾饱和溶液洗涤沉淀？

实验 32　钢铁中镍含量的测定

实验目的

(1) 了解丁二酮肟镍沉淀重量法测定镍的原理和方法。

(2) 掌握重量分析法的基本操作。

实验原理

丁二酮肟是二元弱酸(以 H_2D 表示)，其分子式为 $C_4H_8O_2N_2$，摩尔质量为 116.2 g/mol。研究表明，在氨性溶液中 H_2D 与 Ni^{2+}发生沉淀反应：

$$Ni^{2+} + \begin{array}{c} H_3C - C = NOH \\ \quad\ \ | \\ H_3C - C = NOH \end{array} + 2NH_3 \cdot H_2O =\!=\!=$$

```
                O---H—O
                ↑
H3C — C = N          N = C — CH3
            ˙.     ↙
              Ni                     ↓  + 2NH4+ +2H2O
            ↗    ˙.
H3C — C = N          N = C — CH3
          |          ↓
          O — H---O
```

沉淀经过滤、洗涤，在 120℃下烘干至恒量，可得丁二酮肟镍沉淀的质量，据此可计算 Ni 的质量分数。反应介质是 pH 为 8～9 的氨性溶液，酸度过大或过小都会增加沉淀的溶解度。氨浓度太高，会生成 Ni^{2+}的氨络合物。

丁二酮肟是一种高选择性的有机沉淀剂，它只与 Ni^{2+}、Pd^{2+}、Fe^{2+}生成沉淀。Co^{2+}、Cu^{2+}与其生成水溶性络合物，不仅会消耗 H_2D，且会引起共沉淀现象。当 Co^{2+}，Cu^{2+}含量高时，最好进行二次沉淀或预先分离。此外，由于 Fe^{3+}、Al^{3+}、Cr^{3+}、Ti^{4+} 等离子在氨性溶液中生成氢氧化物沉淀，干扰测定，故溶液在加氨水前，需加入柠檬酸或酒石酸络合剂，使其生成水溶性的络合物。

实验用品

仪器：分析天平，烧杯(500 mL)，量筒(10 mL，50 mL)，表面皿，漏斗，定量滤纸，G4 砂芯坩埚，真空泵，烘箱。

试剂：混合酸($HCl:HNO_3:H_2O$=3：1：2)，酒石酸或柠檬酸溶液(500 g/L)，丁二酮肟(10 g/L，乙醇溶液)，氨水(7 mol/L)，HCl 溶液(6 mol/L)，HNO_3 溶液(2 mol/L)，$AgNO_3$ 溶液(0.1 mol/L)，NH_3-NH_4Cl 洗涤液(100 mL 蒸馏水中加入 1 mL 氨水和 1 g NH_4Cl)，微氨性的酒石酸溶液(20 g/L，pH 为 8～9)，钢铁试样。

实验步骤

准确称取钢铁试样(含 Ni 30～80 mg)于 500 mL 烧杯中，加入 20～40 mL 混合酸，盖上表面皿，低温加热溶解后，煮沸除去氮的氧化物，加入 5～10 mL 酒石酸溶液(500 g/L，每克试样加入 10 mL)。然后，在不断搅动下滴加氨水(7 mol/L)至溶液 pH 为 8～9，此时溶液变为蓝绿色。如有不溶物，应将沉淀过滤，并用热的氨-氯化铵洗涤液洗涤沉淀数次(洗涤液与滤液合并)。滤液用 HCl 溶液(6 mol/L)酸化，用热蒸馏水稀释至约 300 mL，加热至 70～80℃，在不断搅拌下加入丁二酮肟乙醇溶液(10 g/L)沉淀 Ni^{2+}(每毫克 Ni^{2+} 约需 1 mL 10 g/L 丁二酮肟溶液)，最后再多加 20～30 mL。所加试剂总量不超过试液体积的 1/3，以免增大沉淀的溶解量。在不断搅拌下滴加氨水(7 mol/L)，调节溶液的 pH 为 8～9。在 60～70℃下保温 30～40 min。取下，稍冷后，用已恒量的 G4 砂芯坩埚(m_1)进行减压过滤，用微氨性的酒石酸溶液(20 g/L)洗涤烧杯和沉淀 5～8 次，再用温热蒸馏水洗涤沉淀至无 Cl^-为止(检验 Cl^-时，可将滤液用 2 mol/L HNO_3 溶液酸化，用 0.1 mol/L $AgNO_3$ 溶液检验)。将带有沉淀的砂芯坩埚置于 130～150℃烘箱中 1 h，冷却，称量，再烘干，称量，直至恒量(m_2)。平行测定两份试样，将质量 m_1(g)和 m_2(g)记录于表 3.48 中。

实验完毕，砂芯坩埚用稀 HCl 溶液洗涤干净。

数据记录与处理

根据丁二酮肟镍的质量计算试样中的镍含量。

表 3.48 镍含量的测定

	1	2
$m_{试样}$/g		
m_1/g		
m_2/g		
(m_2-m_1)/g		
w_{Ni}/%		
$\overline{w}_{Ni}$/%		
$\lvert d_i \rvert$		
$\overline{d}_r$/%		

思考题

(1) 溶解试样时加入 HNO_3 的作用是什么？

(2) 为了得到纯净的丁二酮肟镍沉淀，应选择和控制好哪些实验条件？

第 4 章　分光光度法及常用分离方法实验

实验 33　邻二氮菲分光光度法测定铁

实验目的

(1) 了解分光光度计的结构和正确的使用方法。

(2) 学习如何选择分光光度分析的实验条件。

(3) 学习吸收曲线、标准曲线的绘制及最大吸收波长的选择。

实验原理

在 pH 为 3～9 的溶液中，邻二氮菲(phen)与 Fe^{2+} 生成稳定的红色络合物，其 $\lg K_f = 21.3$，摩尔吸光系数 $\varepsilon = 1.1\times10^4$ L/(mol·cm)。反应式如下：

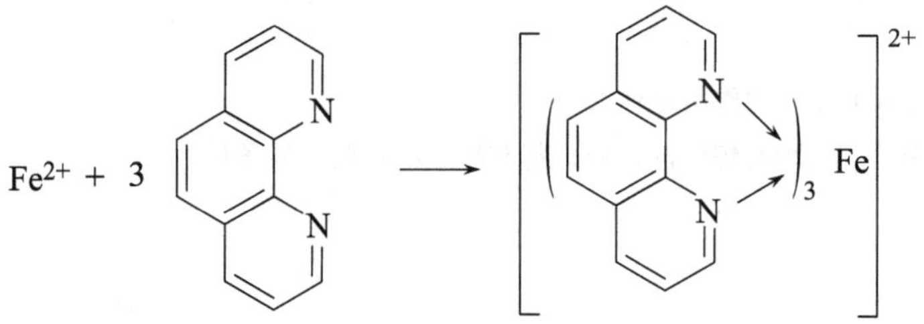

该红色络合物的最大吸收峰在 510 nm 波长处。本方法的选择性很强，相当于含铁量 40 倍的 Sn^{2+}、Al^{3+}、Ca^{2+}、Mg^{2+}、Zn^{2+}、SiO_3^{2-}，20 倍的 Cr^{3+}、Mn^{2+}、V(Ⅴ)、PO_4^{3-}，5 倍的 Co^{2+}、Cu^{2+} 等均不干扰测定。

显色反应受到多种因素的影响，需要通过实验确定其条件。

实验用品

仪器：721(或 722)型分光光度计，吸量管(10 mL)，容量瓶(50 mL)，比色皿(1 cm)，pH 计。

试剂：铁标准储备液[0.1 mg/L，准确称取 0.7020 g $NH_4Fe(SO_4)_2\cdot6H_2O$ 于烧杯中，加入 20 mL H_2SO_4(1∶1)和少量水，溶解后，定量转移至 1 L 容量瓶中，用水稀释至刻度，摇匀]，铁标准溶液(0.001 mol/L，可用铁标准储备液稀释配制)，邻二氮菲溶液(1.5 g/L，避光保存，溶液颜色变暗时即不能使用)，盐酸羟胺水溶液(100 g/L，临用时配制)，乙酸钠溶液(1 mol/L)，NaOH 溶液(0.1 mol/L)，H_2SO_4 溶液(1∶1)，含铁试样。

实验步骤

(1) 实验条件的优化。

(i) 吸收曲线的制作和测量波长的选择。

用吸量管吸取 0 mL、2 mL 铁标准溶液，分别注入两个 50 mL 容量瓶中，各加入 1 mL 盐酸羟胺溶液、2 mL 邻二氮菲、5 mL NaAc，用蒸馏水稀释至刻度，摇匀，放置 10 min。用 1 cm 比色皿，以试剂空白(0 mL 铁标准溶液)为参比，在 400～560 nm，每隔 10 nm 测一次吸光度。以吸光度对波长作图。从吸收曲线上选择测定 Fe 的适宜波长，一般选用最大吸收波长 λ_{max}。

(ii) 显色剂用量的选择。

取 7 个 50 mL 容量瓶，各加入 2 mL 铁标准溶液、1 mL 盐酸羟胺，摇匀。再分别加入 0.2 mL、0.4 mL、0.6 mL、0.8 mL、1.0 mL、2.0 mL、4.0 mL 邻二氮菲和 5 mL NaAc，用蒸馏水稀释至刻度，摇匀，放置 10 min。用 1 cm 比色皿，以蒸馏水为参比溶液，在选择的波长下测定各溶液的吸光度。以吸光度对邻二氮菲溶液体积作图，得出测定铁时显色剂的最佳用量。

(iii) 溶液酸度的选择。

取 7 个 50 mL 容量瓶，分别加入 2 mL 铁标准溶液、1 mL 盐酸羟胺、2 mL 邻二氮菲，摇匀。然后分别加入 0.0 mL、2.0 mL、5.0 mL、10.0 mL、15.0 mL、20.0 mL、30.0 mL NaOH 溶液(0.1 mol/L)，用水稀释至刻度，摇匀，放置 10 min。用 1 cm 比色皿，以蒸馏水为参比溶液，在选择的波长下测定各溶液的吸光度。同时，用 pH 计测量各溶液的 pH。以吸光度对 pH 作图，得出测定铁的适宜酸度范围。

(iv) 显色时间。

在 50 mL 容量瓶中加入 2 mL 铁标准溶液、1 mL 盐酸羟胺溶液，摇匀。再加入 2 mL 邻二氮菲、5 mL NaAc，用水稀释至刻度，摇匀。立即用 1 cm 比色皿，以蒸馏水为参比溶液，在选择的波长下测量吸光度。然后依次测量溶液放置 5 min、10 min、30 min、60 min、120 min 后的吸光度。以吸光度对显色时间作图，得出铁与邻二氮菲显色反应完全所需的适宜时间。

(v) 邻二氮菲与铁的摩尔比的测定。

取 8 个 50 mL 容量瓶，吸取 0.001 mol/L 铁标准溶液 10 mL 于各容量瓶中，各加 1 mL 盐酸羟胺溶液、5 mL NaAc。然后依次加 0.5 mL、1.0 mL、2.0 mL、2.5 mL、3.0 mL、3.5 mL、4.0 mL、5.0 mL 邻二氮菲，用水稀释至刻度，摇匀，放置 10 min。然后用 1 cm 比色皿，在选择的波长下，以蒸馏水为参比溶液，测定各溶液的吸光度。最后以吸光度对邻二氮菲与铁的浓度比(c_{Phen}/c_{Fe})作图，根据曲线上前后两部分延长线的交点位置确定 Fe^{2+} 与邻二氮菲反应的络合比。

(2) 铁含量的测定。

(i) 标准曲线的制作。

在 6 个 50 mL 容量瓶中，用吸量管分别加入 0.0 mL、2.0 mL、4.0 mL、6.0 mL、8.0 mL、10.0 mL 铁标准溶液(0.001 mol/L)，分别加入 1 mL 盐酸羟胺、2 mL 邻二氮菲、5 mL NaAc，每加一种试剂后摇匀。然后用水稀释至刻度，摇匀，放置 10 min。用 1 cm 比色皿，以试剂空白(0.0 mL 铁标准溶液)为参比溶液，在选择的波长下，测量各溶液的吸光度。以吸光度对铁含量作图，绘制标准曲线，计算 Fe^{2+}-phen 络合物的摩尔吸光系数。

(ii) 试样中铁含量的测定。

含铁试样溶液按步骤(i)显色后，在相同条件下测量吸光度(A)，根据标准曲线求出试样中铁的含量。

数据记录与处理

绘制各种实验条件曲线、标准曲线，并计算试样中铁的含量。

思考题

(1) 制作标准曲线和进行其他条件实验时，加入试剂的顺序能否任意改变？为什么？

(2) 吸收曲线与标准曲线有何区别？

(3) 本实验中盐酸羟胺、乙酸钠的作用各是什么？

(4) 如何用分光光度法测定水样中的总铁和亚铁离子的含量？试拟出简单步骤。

实验 34　食品中 NO_2^- 含量的测定

实验目的

(1) 掌握分光光度计的使用方法。

(2) 掌握测定食品中 NO_2^- 含量的方法。

实验原理

亚硝酸盐作为一种食品添加剂，能够保持腌制品的色香味，并具有一定的防腐性。但同时也具有较强的致癌作用，过量食用会对人体产生危害。

在弱酸性溶液中，亚硝酸盐与对氨基苯磺酸发生重氮化反应，生成的重氮化合物与盐酸萘乙二胺偶联成紫红色的偶氮染料，可用分光光度法测定。有关反应式如下:

$$NO_2^- + 2H^+ + H_2N\text{—}C_6H_4\text{—}SO_3H \longrightarrow N{\equiv}\overset{+}{N}\text{—}C_6H_4\text{—}SO_3H + 2H_2O$$

$$N{\equiv}\overset{+}{N}\text{—}C_6H_4\text{—}SO_3H + C_{10}H_7\text{—}NHCH_2CH_2NH_2\cdot HCl \longrightarrow$$

$$HO_3S\text{—}C_6H_4\text{—}N{=}N\text{—}C_{10}H_6\text{—}NHCH_2CH_2NH_2\cdot HCl$$

实验用品

仪器：721 型分光光度计，多用食品粉碎机，容量瓶(250 mL)，滤纸，漏斗，比色皿(2 cm)。

试剂：硼砂饱和溶液(称取 25 g $Na_2B_4O_7\cdot 10H_2O$ 溶于 500 mL 热水中)，硫酸锌溶液

(1.0 mol/L，称取 150 g $ZnSO_4 \cdot 7H_2O$ 溶于 500 mL 水中)，对氨基苯磺酸溶液(4 g/L，称取 0.4 g 对氨基苯磺酸溶于 200 g/L 盐酸中配成 100 mL 溶液，避光保存)，盐酸萘乙二胺溶液(2 g/L，称取 0.2 g 盐酸萘乙二胺溶于 100 mL 水中，避光保存)，$NaNO_2$ 标准储备液(准确称取 0.1000 g 干燥 24 h 的分析纯 $NaNO_2$，溶解后定量转入 500 mL 容量瓶中，加水稀释至刻度并摇匀)，$NaNO_2$ 操作液(1 μg/mL，临用时由上述储备液配制)，活性炭。

实验步骤

(1) 试样预处理。

称取 5 g 绞碎均匀的腌制肉制品试样于 50 mL 烧杯中，加入 12.5 mL 硼砂饱和溶液搅拌均匀，然后用 150～200 mL 70℃以上的热水将烧杯中的试样全部洗入 250 mL 容量瓶中，并置于沸水浴中加热 15 min，取出。在轻轻摇动下滴加 2.5 mL $ZnSO_4$ 溶液，以沉淀蛋白质。冷却至室温后，加水稀释至刻度，摇匀。放置 10 min。弃去上层脂肪，清液用滤纸或脱脂棉过滤，弃去最初 10 mL 滤液，测定用滤液应为无色透明。

(2) 样品分析。

(i) 标准曲线的绘制。

准确移取 $NaNO_2$ 操作液(1 μg/mL) 0.0 mL、0.4 mL、0.8 mL、1.2 mL、1.6 mL、2.0 mL 分别置于 6 个 50 mL 容量瓶中，各加 30 mL 水，然后分别加入 2 mL 对氨基苯磺酸溶液，摇匀。静置 3 min 后，再分别加入 1 mL 盐酸萘乙二胺溶液，加水稀释至刻度，摇匀。放置 15 min，用 2 cm 比色皿，以试剂空白为参比溶液，于波长 540 nm 处测定各试液的吸光度。

(ii) 试样的测定。

准确移取 40 mL 经过处理的试样滤液于 50 mL 容量瓶中，按“标准曲线的绘制”操作(不需稀释)。根据测得的吸光度，从标准曲线上查出相应 $NaNO_2$ 的浓度。

数据记录与处理

以 $NaNO_2$ 溶液的加入量为横坐标，相应的吸光度为纵坐标绘制标准曲线。根据标准曲线，计算试样中 $NaNO_2$ 的质量分数(以 mg/kg 表示)。

注意事项

(1) 亚硝酸盐容易氧化为硝酸盐，处理试样时加热的时间和温度均要注意控制。另外，配制的标准储备液不宜久存。

(2) 本法测量中不包括试样中硝酸盐的含量。

思考题

(1) 亚硝酸盐作为一种食品添加剂，具有哪些特点？能否找到一种优于亚硝酸盐的替代品？

(2) 盛接滤液时，为什么要弃去最初的 10 mL 滤液？

实验 35　土壤中有效磷的测定

实验目的

(1) 了解光度法测定土壤中有效磷的原理和方法。

(2) 掌握分光光度计的使用方法。

实验原理

土壤中有效磷含量是指能被当季作物吸收的磷量。常用化学法测定土壤中的有效磷，即用浸提剂(根据土壤的性质选择)提取土壤中的一部分有效磷。酸性土壤中，磷酸铁和磷酸铝形态的有效磷可用酸性氟化铵提取，形成氟铝化铵和氟铁化铵络合物，少量的钙离子形成氟化钙沉淀，磷酸根离子被提取到溶液中。石灰性土壤则采用碳酸氢钠溶液浸取。在含磷的溶液中加入钼酸铵，在一定酸度条件下，溶液中的磷酸与钼酸络合形成黄色的磷钼杂合酸(磷钼黄)。

在适宜的试剂浓度下，加入适当的还原剂($SnCl_2$或抗坏血酸)，使磷钼酸中的一部分Mo(Ⅵ)还原为Mo(Ⅴ)，生成磷钼蓝(磷钼杂多蓝$H_3PO_4 \cdot 10MoO_3 \cdot Mo_2O_5$或$H_3PO_4 \cdot 8MoO_3 \cdot 2Mo_2O_5$)。在一定的浓度范围内，蓝色的深浅与磷含量成正比，这是钼蓝比色法的基础。

实验用品

仪器：721(或722)型分光光度计，容量瓶(25 mL)，带塞比色管(50 mL)，吸量管(5 mL)，量筒(50 mL)，振荡机，漏斗，滤纸。

试剂：HCl溶液(0.5 mol/L)，NH_4F溶液(1 mol/L)，提取剂(分别移取15 mL 1 mol/L NH_4F溶液和25 mL 0.5 mol/L HCl溶液，加入460 mL蒸馏水中，配制成0.03 mol/L NH_4F-0.025 mol/L HCl溶液)，H_3BO_3溶液(100 g/L)，15 g/L 钼酸铵-3.5 mol/L 盐酸溶液(15 g钼酸铵溶于300 mL蒸馏水中，加热至60℃左右，如有沉淀，将溶液过滤，待溶液冷却后，慢慢加入350 mL 10 mol/L HCl溶液，并用玻璃棒迅速搅动，待溶液冷却至室温，用蒸馏水稀释至1 L，充分摇匀，存于棕色瓶中，放置时间不得超过两个月)，氯化亚锡溶液(25 g/L，2.5 g氯化亚锡溶于10 mL浓HCl中，溶解后加入90 mL蒸馏水，混合均匀置于棕色瓶中，此溶液现配现用)，磷标准溶液[50 μg/mL，将0.2195 g KH_2PO_4(AR，105℃烘干)溶于400 mL蒸馏水中，加5 mL浓H_2SO_4防止溶液长霉菌，转入1 L容量瓶中，加蒸馏水稀释至刻度，摇匀。准确移取25.00 mL上述溶液于250 mL容量瓶中，稀释至刻度，摇匀，即为50 μg/mL磷标准溶液，此溶液不宜久存]。

实验步骤

(1) 土壤样品预处理。

称取1 g(精确至0.01 g)风干土壤样品，放入50 mL带塞比色管中，加20 mL 0.03 mol/L NH_4F-0.025 mol/L HCl溶液，稍摇匀，立即放在振荡机上振荡30 min。用无磷干滤纸过滤，滤液盛接于盛有15滴100 g/L H_3BO_3溶液的50 mL锥形瓶中，摇动瓶内溶液(加H_3BO_3防止F^-对显色的干扰和腐蚀玻璃仪器)。

(2) 土壤中有效磷的测定。

准确移取 5～10 mL 上述土壤滤液于 25 mL 容量瓶中，用吸量管加入 5 mL 15 g/L 钼酸铵盐酸溶液，摇匀，加蒸馏水至近瓶颈刻度，滴加 3 滴 25 g/L 氯化亚锡后，再用水稀释至刻度，充分摇匀。显色 15 min 后，在分光光度计上，以试剂空白为参比溶液，用 1 cm 比色皿在 680 nm 处测其吸光度值。

(3) 标准曲线的绘制。

分别准确移取 5 μg/mL 磷标准溶液 0.0 mL、1.0 mL、2.0 mL、3.0 mL、4.0 mL、5.0 mL 于 6 个 25 mL 容量瓶中，加入 5～10 mL 0.03 mol/L NH_4F-0.025 mol/L HCl 溶液(按所取滤液体积而定)，用吸量管加 5 mL 钼酸铵盐酸溶液，加蒸馏水至近瓶颈刻度，并滴加 3 滴 25 g/L 氯化亚锡，摇动后，至溶液有深蓝色出现，用水稀释至刻度，摇匀，放置 15 min，在与土壤样品相同的实验条件下测其吸光度。

数据记录与处理

以磷的质量(μg)为横坐标，相应的吸光度为纵坐标绘制标准曲线，并从标准曲线上查出土壤样品中磷的含量。

注意事项

用氯化亚锡作还原剂生成磷钼盐，溶液的颜色不够稳定，必须严格控制比色时间，一般在显色后的 15～20 min 颜色较为稳定。因此，显色后应在 15 min 后立即比色，并在 5 min 内完成比色操作。

思考题

(1) 试述本实验测定磷的基本原理。

(2) 氯化亚锡溶液放置过久，对实验有什么影响？

实验 36　微波辅助萃取-分光光度法测定竹叶中总黄酮

实验目的

(1) 了解微波辅助萃取的原理和方法。

(2) 掌握微波辅助萃取用于实际试样分离提取的操作技术。

实验原理

微波是频率为 300～300000 MHz，即波长为 1 mm 至 1 m 的电磁波。微波辅助萃取(microwave-assisted extraction，MAE)是在传统有机溶剂萃取的基础上发展起来的一种新型分离提取技术。通过选用适当溶剂，根据化合物不同的物理化学性质及对微波吸收能力的差异，利用微波能加快目标化合物与基体分离而溶出到溶剂中的速率，从而提高从基体中提取目标化合物的提取效率。

微波辐射是一个“体加热”过程，用于天然产物试样提取分离时，细胞内温度迅速上升，使得细胞内压力急剧增大，导致细胞破裂，有利于在回流过程中提取剂进入细胞内，从而使得包含在细胞内的有效成分快速释放出来。因此，微波辅助萃取具有快速、节能、省溶剂、环境友好等特点。

近年来对竹子的研究发现，竹叶中含有大量对人体有益的活性物质，包括黄酮类、活性多糖、特种氨基酸等，其中主要是黄酮苷等酚类化合物。现代科学研究表明，竹叶黄酮具有优良的抗自由基、抗氧化、抗衰老、抗菌、抗病毒及保护心脑血管等作用。本实验采用微波辅助萃取技术从竹叶中提取分离黄酮类化合物，并采用分光光度法测定总黄酮的含量。

实验用品

仪器：圆底烧瓶(50 mL)，容量瓶(50 mL)，移液管(1 mL，2 mL，5 mL)，微波萃取仪，紫外-可见分光光度计，离心机，分析天平。

试剂：无水乙醇，芦丁标准溶液(300 μg/mL，准确称取 75 mg 芦丁标准品，用 60%乙醇溶解并定容至 250 mL，摇匀)，$NaNO_2$ 溶液(50 g/L)，$Al(NO_3)_3$ 溶液(100 g/L)，NaOH 溶液(1 mol/L)，竹叶粉末(剪下竹叶，用自来水和去离子水先后洗净，放于烘箱内，60℃烘焙 7 h。待竹叶冷至室温后，放入搅拌机中搅碎，并过 20 目筛)。

实验步骤

(1) 微波辅助萃取竹叶中黄酮。

称取 1.00 g 竹叶粉末于 50 mL 圆底烧瓶中，加入 15 mL 20%乙醇溶液作为萃取剂，摇匀，放入磁力搅拌子，设定转速为 500 r/min，75℃下在微波萃取仪中萃取 25 min。萃取结束后过滤，滤液用石油醚除去叶绿素和可溶性脂类，转移至 50 mL 容量瓶中，用 20%乙醇溶液定容至刻度。

(2) 标准曲线的绘制。

分别取 0.5 mL、1.0 mL、1.5 mL、2.0 mL、2.5 mL、3.0 mL 300 μg/mL 芦丁标准溶液于 6 个 10 mL 比色管中，加入去离子水至 5.0 mL。加入 0.50 mL 50 g/L $NaNO_2$ 溶液，摇匀，放置 6 min；加 0.50 mL 100 g/L $Al(NO_3)_3$ 溶液，摇匀，放置 6 min；加 4.0 mL 1 mol/L NaOH 溶液，加去离子水定容至 10 mL，摇匀。放置 10 min 后离心分离 5 min，注入 1 cm 比色皿，以试剂空白作为参比溶液，于 510 nm 波长下测定吸光度，将数据记录于表 4.1 中。

(3) 试样测定。

取 5.00 mL 竹叶的微波提取液于 10 mL 比色管中，按芦丁标准溶液的测定方法进行总黄酮的测定，根据标准曲线计算出总黄酮的含量(以芦丁的质量分数表示)。将数据记录于表 4.1 中。

数据记录与处理

表 4.1　标准溶液的吸光度及试样中总黄酮含量的测定

	芦丁的质量浓度/(mg/L)	吸光度	试样中总黄酮的含量/%
1			—
2			—
3			—

续表

	芦丁的质量浓度/(mg/L)	吸光度	试样中总黄酮的含量/%
4			—
5			—
6			—
试样溶液			

思考题

(1) 微波辅助萃取与普通溶剂萃取相比具有什么优势？

(2) 微波加热的方式有什么特点？影响微波辅助萃取的因素主要有哪些？

实验 37　萃取分离-分光光度法测定环境水样中微量铅

实验目的

(1) 了解双硫腙萃取分光光度法测定环境水样中铅的原理和方法。

(2) 掌握萃取分离的基本操作。

实验原理

铅是一种积累性毒物，易被肠胃吸收，通过血液影响酶和细胞的新陈代谢。过量铅的摄入将严重影响人体健康，引起贫血、神经机能失调和肾损伤等。我国《生活饮用水卫生标准》规定，生活饮用水中含铅量不能超过 0.01 mg/L。因此，铅在环境水样中的含量是环境监测的一个重要指标。

双硫腙萃取分光光度法是现行国家环境标准监测方法中测定水中铅的方法之一。方法的主要原理是：在 pH 为 8.5～9.5 的氨性柠檬酸盐-氯化物-盐酸羟胺的还原性介质中，铅与双硫腙形成淡红色双硫腙螯合物：

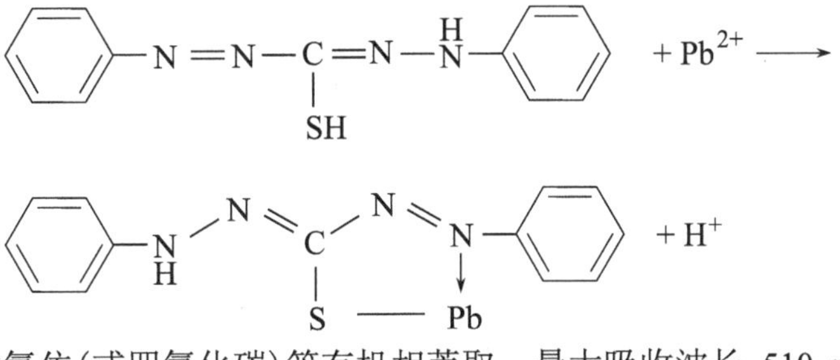

$$C_6H_5-N=N-\underset{\underset{SH}{|}}{C}=N-\overset{H}{N}-C_6H_5 + Pb^{2+} \longrightarrow$$

$$C_6H_5-\underset{H}{N}-N=\underset{\underset{S}{|}}{C}-N=\underset{\underset{Pb}{\downarrow}}{N}-C_6H_5 + H^+ \quad (S—Pb)$$

该螯合物可被氯仿(或四氯化碳)等有机相萃取，最大吸收波长 510 nm，摩尔吸光系数 6.7×10^4 L/(mol·cm)。试样溶液中加入盐酸羟胺，还原 Fe^{3+} 及可能存在的其他氧化性物质，防止双硫腙被氧化；加入氰化物掩蔽 Ag^+、Hg^{2+}、Cu^{2+}、Zn^{2+}、Cd^{2+}、Ni^{2+}、Co^{2+} 等；加入柠檬酸盐络合 Al^{3+}、Cr^{3+}、Fe^{3+}、Ca^{2+}、Mg^{2+} 等，防止它们在碱性溶液中水解沉淀。双硫腙萃取分光光度法在实验过程中具有萃取、分离、富集等步骤，因此选择性和灵敏度较高，适用于测定地表水和废水中微量铅。

实验用品

仪器：容量瓶(250 mL)，移液管(1 mL，2 mL，5 mL)，分光光度计，分液漏斗(250 mL)。

试剂：铅标准溶液[2.0 μg/mL，准确称取 0.1599 g $Pb(NO_3)_2$(纯度≥ 99.5%)溶于约 200 mL 去离子水中，加入 10 mL 浓 HNO_3，移入 1000 mL 容量瓶，用蒸馏水稀释至刻度，此溶液含铅 100.0 μg/mL。移取此溶液 10.00 mL 置于 500 mL 容量瓶中，用蒸馏水稀释至刻度，摇匀]，双硫腙储备液(0.1 g/L，称取 0.1000 g 双硫腙溶于 1000 mL 氯仿中，储于棕色瓶，0～4℃保存)，双硫腙工作液(0.04 g/L，移取 100 mL 双硫腙储备液置于 250 mL 容量瓶中，用氯仿稀释至刻度)，双硫腙专用液(将 250 mg 双硫腙溶于 250 mL 氯仿中，此溶液不必纯化，专用于萃取提纯试剂)，柠檬酸-氰化钾还原性氨性溶液[将 100 g 柠檬酸氢二铵、5 g 无水 Na_2SO_3、2.5 g 盐酸羟胺、10 g KCN(剧毒！)溶于蒸馏水，用蒸馏水稀释至 250 mL，加入 500 mL 氨水混合]。

实验步骤

(1) 水样预处理。

洁净程度高的水(如不含悬浮物的地下水、清洁地面水)可直接测定。其他情况预处理过程如下：

(i) 浑浊的地面水：取 250 mL 水样加入 2.5 mL 浓 HNO_3，于电热板上微沸消解 10 min，冷却后用快速滤纸滤入 250 mL 容量瓶，滤纸用 0.03 mol/L HNO_3 洗涤数次，并稀释至刻度。

(ii) 含悬浮物和有机物较多的水样：取 200 mL 水样加入 10 mL 浓 HNO_3 煮沸消解至 10 mL 左右，稍冷却，补加 10 mL 浓 HNO_3 和 4 mL 浓 $HClO_4$，继续消解蒸至近干。冷却后用 0.03 mol/L HNO_3 温热溶解残渣，冷却后用快速滤纸滤入 200 mL 容量瓶，用 0.03 mol/L HNO_3 洗涤滤纸并定容。

(2) 标准曲线的绘制。

在 8 个 250 mL 分液漏斗中分别加入 0.0 mL、0.50 mL、1.00 mL、5.00 mL、7.50 mL、10.00 mL、12.50 mL、15.00 mL 铅标准溶液，补加去离子水至 100 mL，加入 10 mL 3 mol/L HNO_3 和 50 mL 柠檬酸盐-氰化钾还原性氨性溶液，混匀。再加入 10.00 mL 双硫腙工作液，剧烈振荡 30 s，静置分层。在分液漏斗的颈管内塞入一团无铅脱脂棉，放出下层有机相，弃去前面 1～2 mL 流出液后，将有机相注入 1 cm 比色皿，以氯仿为参比溶液，在 510 nm 处测量吸光度，将数据记录于表 4.2 中。

(3) 试样分析。

准确量取适量按实验步骤(1)预处理后的环境水样于 250 mL 分液漏斗中，用去离子水稀释至 100 mL，按标准曲线测定步骤进行测定。将数据记录于表 4.2 中。

表 4.2　标准曲线的绘制及水样中铅含量的测定

	铅含量/μg	吸光度	试样中铅的质量浓度/(μg/L)
1			—
2			—
3			—

续表

	铅含量/μg	吸光度	试样中铅的质量浓度/(μg/L)
4			—
5			—
6			—
7			—
8			—
水样			

数据记录与处理

以铅含量(μg)为横坐标，吸光度为纵坐标绘制标准曲线。由标准曲线计算得到铅含量(μg)，根据水样的体积计算环境水样中铅的质量浓度(μg/L)。

思考题

(1) 为什么用分光光度法测定环境水样中的微量铅要采用萃取分离，而测定矿样中的铅可以不用？

(2) 双硫腙工作液是否需要准确加入？为什么？

(3) 水样预处理的目的是什么？

实验 38　纸色谱法分离和鉴定氨基酸

实验目的

(1) 了解纸色谱法分离、鉴定氨基酸的原理。

(2) 掌握纸色谱法的操作技术和比移值的测定方法。

(3) 学习根据组分的比移值鉴别未知试样中的组分。

实验原理

纸色谱法是以滤纸作为支撑体的分离方法，利用滤纸吸收的水分作固定相，有机溶剂作流动相。流动相由于毛细作用自下而上移动，样品中的各组分在两相中不断进行分配。由于它们的分配系数不同，不同溶质随流动相移动的速度不等，因而形成与原点距离不同的层析点，达到分离的目的。各组分在滤纸上移动的情况用比移值(R_f)表示，其定义为

$$R_f = \frac{\text{斑点中心至起始线的距离}}{\text{溶剂前沿至起始线的距离}} = \frac{a}{b}$$

在一定条件下，R_f值是物质的特征值，故可根据R_f进行定性分析。但R_f的影响因素较多(如固定相与流动相的性质、温度等)，需用各组分相应的标准试样做对照实验。

本实验进行胱氨酸、甘氨酸和酪氨酸的分离和鉴定(其R_f值依次增大)。当氨基酸混合试样在滤纸上点样后，试样溶于固定相中。将滤纸末端浸入展开剂(正丁醇、冰醋酸和水的混合物)，由于毛细作用，流动相沿着滤纸上行，试样中各种氨基酸组分在固定相和流动相中不断进行分配。由于它们的分配系数不同，不同溶质随流动相移动的速度也不同，形成距原点距

离不等的斑点。氨基酸本身无色，鉴定氨基酸常用茚三酮显色剂显色，即层析后在纸上喷洒显色剂茚三酮，斑点呈红紫色。该法比较灵敏，可用于检测微克级的氨基酸。

实验用品

仪器：层析筒(150 mm × 300 mm，$\Phi \times h$)，毛细管，喷雾器，中速色谱滤纸(裁成 90 mm × 240 mm)。

试剂：展开剂(正丁醇 ∶ 冰醋酸 ∶ 水=4 ∶ 1 ∶ 2)，氨基酸标准溶液(胱氨酸、甘氨酸、酪氨酸均为 5 g/L 水溶液)，氨基酸混合溶液(由上述三种氨基酸标准溶液等量混合而成)，茚三酮(2 g/L，正丁醇溶液)。

实验步骤

(1) 点样。

于层析纸一端 3 cm 处轻轻画一条水平横线，在横线上画四个点作为原点，原点间距离为 2 cm。分别用毛细管将三种氨基酸标准溶液及氨基酸混合试液依次点在四个原点处，斑点直径为 2～2.5 mm。在滤纸另一端 2 cm 处穿一根棉线，将纸条晾干。

(2) 展开。

在干燥的层析筒中加入 60 mL 展开剂，把点好样的纸条挂在层析筒盖上，层析纸下端浸入展开剂约 0.5 cm，但原点必须离开液面。盖上层析筒，当溶剂前沿上升到距滤纸上端 2～3 cm，取出层析纸，用铅笔画出溶剂前沿位置。

(3) 显色。

将展开后的层析纸在空气中晾干，用喷雾器将茚三酮溶液喷洒在滤纸上，稍干后，放入烘箱中(90℃左右)烘 3～5 min，滤纸上即显出红紫色斑点。

(4) 测量。

用铅笔标出斑点的范围，找出斑点的中心，用直尺量出各斑点中心到起始线的距离 a，再量出溶剂前沿至起始线的距离 b。将数据记录于表 4.3 中。

数据记录与处理

根据 a、b 值分别求出混合物与标准样品斑点的 R_f。通过对比 R_f，对混合物试样的组分进行定性分析。

表 4.3 氨基酸混合试样分析

	a	b	R_f
胱氨酸			
甘氨酸			
酪氨酸			
氨基酸混合试样			

注意事项

(1) 实验时纸条应挂平直，原点应离开液面，纸条应与展开剂接触。

(2) 注意指纹含有一定量的氨基酸(皮肤会分泌氨基酸)，不能用手指直接接触滤纸，要用镊子钳夹滤纸边。

思考题

(1) 利用纸色谱法分离氨基酸的原理是什么？

(2) 用手指直接接触滤纸，对实验结果有何影响？

(3) 若原点浸入展开剂中，实验结果会怎样？

(4) 纸色谱法中为什么常采用标准样品鉴定未知样品？

实验 39　纸色谱法分离食用色素

实验目的

(1) 了解纸色谱法分离食用色素的原理。

(2) 掌握样品中色素的富集及测定方法。

实验原理

纸色谱法原理同实验 38。本实验是饮料中合成色素的分离。样品处理后，在酸性条件下，用聚酰胺吸附人工合成色素，使其与蛋白质、淀粉、脂肪、天然色素分离，然后在碱性条件下，用适当的解吸溶液使色素解吸出来。由于不同色素的分配系数不同，R_f就不同，可对其分离鉴别。

实验用品

仪器：烧杯(100 mL)，分析天平，层析筒(150 mm × 300 mm，$\Phi \times h$)，砂芯漏斗(G2 或 G3)，层析纸(10 cm × 27.5 cm，$\omega \times h$)，毛细管(直径 1 mm)。

试剂：色素标准溶液(5 g/L，胭脂红、柠檬黄、日落黄)，展开剂(正丁醇 ∶ 无水乙醇 ∶ 氨水=6 ∶ 2 ∶ 3)，柠檬酸溶液(200 g/L)，聚酰胺粉(尼龙 6 过 200 目筛，预先在 105℃温度下活化 1 h)，丙酮-氨水溶液(90 mL 丙酮与 100 mL 浓氨水混合均匀)。

实验步骤

(1) 样品处理。

取 50 mL 除去 CO_2 的橙汁饮料于 100 mL 烧杯中，用柠檬酸溶液调 pH ≈ 4。

(2) 吸附分离。

称取 0.5～1.0 g 聚酰胺于 100 mL 烧杯中，加少量水调成均匀糊状，倒入上述已处理的温度为 70℃的样品溶液中，充分搅拌，使样品溶液中色素完全被吸附(聚酰胺粉不足可补加)。将沉淀物全部转入砂芯漏斗中抽滤，用 70℃蒸馏水洗涤沉淀物，洗涤时充分搅拌，再用 20 mL

丙酮溶液分两次洗涤沉淀物，除去样品中的油脂等物。再用 200 mL 70℃蒸馏水洗涤沉淀，直至洗下的水与原来水的 pH 相同为止。洗涤过程中必须充分搅拌。

用约 30 mL 丙酮-氨水溶液多次解吸色素。将解吸色素置于小烧杯中，用柠檬酸调节至 pH≈6，再在水浴上蒸发浓缩至 5 mL 留作点样用。

(i) 点样。

在层析纸下端 2.5 cm 处用铅笔画一横线，在线上等距离画上 1、2、3、4 四个等距离的点。1、2、3 号分别用毛细管将胭脂红、柠檬黄和日落黄色素标准溶液点出直径为 2 mm 的扩散原点；在 4 号点上样品溶液，每点完一次须用电吹风吹干，再在原位置上重新点。

(ii) 展开分离。

将点好样的滤纸晾干后，用挂钩悬挂在层析筒盖上，放入已盛有展开剂的层析筒中，滤纸应挂平直，原点应离开液面 1 cm，保持温度 20℃，密封层析筒，按上行法展开。当展开剂前沿滤纸上升到 12 cm 处时，将滤纸取出，在空气中自然晾干。量出各斑心的中点到原点中心的距离。将数据记录于表 4.4 中。

数据记录与处理

根据每个斑点的移动距离计算 R_f 值。若 R_f 值相同，色泽相似，表示被测色素与标准色素为同一色素。

表 4.4　食用色素分析

	a	b	R_f
胭脂红			
柠檬黄			
日落黄			
样品分析			

注意事项

(1) 聚酰胺是高分子化合物，在酸性介质中才能吸附酸性色素。为防止色素分解，要保持酸性条件。

(2) 分子中酰胺链能与色素中磺酸基以氢键的形式结合，所以吸附时也要求一定的温度与时间。

思考题

(1) 纸色谱法分离合成色素时，流动相和固定相各是什么？

(2) 洗涤聚酰胺时要注意哪些方面？为什么？

(3) 处理样品所得的溶液，为什么要调 pH≈4？

第5章　综合设计实验

本章实验的目的主要是使学生进一步熟悉和巩固称量、移液、滴定等有关知识和实验操作技能，培养独立操作、独立分析问题和解决问题的能力。学生应根据所选定的实验题目，查阅有关的参考资料，设计分析方案(包括实验原理、试剂配制、标准溶液的配制和标定、指示剂、仪器、取样量、固体试样的溶样方法、具体的分析步骤及数据处理等)。分析方案经教师审阅后，进行实验并写出实验报告。

5.1　酸碱滴定设计实验

实验40　氟硅酸钾法测定SiO_2

合金试样经硝酸和氢氟酸分解，使硅转化为硅酸，然后在大量钾离子存在下，形成氟硅酸钾沉淀(硅酸盐试样经KOH熔融分解后转化为可溶性硅酸盐，它在强酸介质中与KF形成难溶的氟硅酸钾)：

$$2K^+ + SiO_3^{2-} + 6F^- + 6H^+ \Longrightarrow K_2SiF_6\downarrow + 3H_2O$$

沉淀溶解度较大，沉淀时需加入固体KCl降低其溶解度。经过滤、洗涤除去大部分游离酸，以酚酞为指示剂，用氢氧化钠标准溶液中和剩余游离酸。将生成的K_2SiF_6沉淀滤出，加入沸水使沉淀水解，再用氢氧化钠标准溶液滴定水解产生的氢氟酸。反应式如下：

$$K_2SiF_6 + 3H_2O \Longrightarrow 2KF + H_2SiO_3 + 4HF$$

由于生成的HF对玻璃有腐蚀作用，因此操作必须在塑料容器中进行。

参考文献

傅晓珊. 2015. 氟硅酸钾容量法测定合金中硅含量的研究. 福建分析测试, 24(5): 41-42
李环亭，董艳艳，刘晓毅，等. 2011. 氟硅酸钾-酸碱滴定法测定硅质耐火材料中SiO_2. 冶金分析, 31(2): 67-70

实验41　饼干中$NaHCO_3$和Na_2CO_3含量的测定

在制作饼干时，为了增加饼干的酥脆口感，通常会加入Na_2CO_3或$NaHCO_3$。欲测定同一份试样中各组分的含量，可用HCl标准溶液滴定。根据滴定过程中pH变化的情况，选用酚酞和甲基橙为指示剂，称为“双指示剂法”。

双指示剂法中，一般先用酚酞，后用甲基橙指示剂。由于以酚酞作指示剂时从微红色到无色的变化不敏锐，因此也常选用甲酚红-百里酚蓝混合指示剂。甲酚红的变色范围为6.7(黄色)～8.4(红色)，百里酚蓝的变色范围为8.0(黄色)～9.6(蓝色)，混合后的变色点是8.3，酸色为黄色，碱色为紫色，混合指示剂变色敏锐。用 HCl 标准溶液滴定试液由紫色变为粉红色，即为终点。

参考文献

黄伟坤. 1997. 食品检验与分析. 北京: 中国轻工业出版社, 592

实验 42　矿渣中三氧化二硼的测定

二氧化硫分解硼镁矿制取硼酸的方法具有工艺简单、流程短、分解率和回收率较高等优点，是硼酸生产的一种新工艺。对于矿渣的综合利用，其中一条主要途径是氧化后生产含硼复合肥料。因此，测定矿渣中的三氧化二硼具有一定的实际意义。

硼酸是多元酸，但酸性极弱(K_a=5.8 × 10^{-10})，不能直接用碱滴定。但硼酸根能与甘油、甘露醇等形成解离度远大于硼酸的羟基络合物，使硼酸变为中强酸。反应式如下：

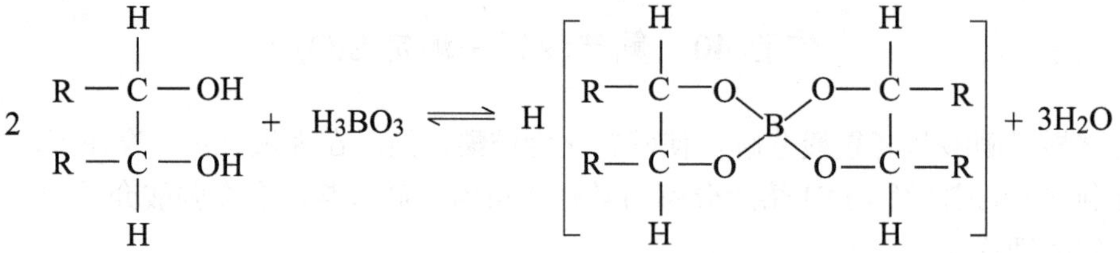

$$2\ \begin{array}{c} \text{H}\\ | \\ \text{R}-\text{C}-\text{OH}\\ |\\ \text{R}-\text{C}-\text{OH}\\ |\\ \text{H}\end{array} + H_3BO_3 \rightleftharpoons \text{H}\left[\begin{array}{ccccccc} & \text{H} & & & & \text{H} & \\ & | & & & & | & \\ \text{R}- & \text{C} & -\text{O} & & \text{O}- & \text{C} & -\text{R}\\ & | & & \text{B} & & | & \\ \text{R}- & \text{C} & -\text{O} & & \text{O}- & \text{C} & -\text{R}\\ & | & & & & | & \\ & \text{H} & & & & \text{H} & \end{array}\right] + 3H_2O$$

该络合物的 pK_a＝4.26，可用 NaOH 标准溶液准确滴定。滴定反应为

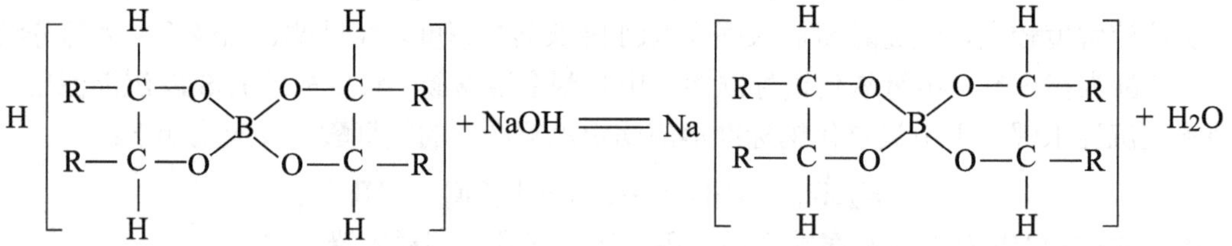

$$\text{H}\left[\begin{array}{ccccccc} & \text{H} & & & & \text{H} & \\ & | & & & & | & \\ \text{R}- & \text{C} & -\text{O} & & \text{O}- & \text{C} & -\text{R}\\ & | & & \text{B} & & | & \\ \text{R}- & \text{C} & -\text{O} & & \text{O}- & \text{C} & -\text{R}\\ & | & & & & | & \\ & \text{H} & & & & \text{H} & \end{array}\right] + \text{NaOH} = \text{Na}\left[\begin{array}{ccccccc} & \text{H} & & & & \text{H} & \\ & | & & & & | & \\ \text{R}- & \text{C} & -\text{O} & & \text{O}- & \text{C} & -\text{R}\\ & | & & \text{B} & & | & \\ \text{R}- & \text{C} & -\text{O} & & \text{O}- & \text{C} & -\text{R}\\ & | & & & & | & \\ & \text{H} & & & & \text{H} & \end{array}\right] + H_2O$$

此反应是等摩尔进行，化学计量点的 pH 为 9.2 左右，可选酚酞或百里酚蓝为指示剂，从而测定试样中的三氧化二硼。

参考文献

程相春. 2011. 硼酸测定中强化试剂效果比较. 化学工程师, 5: 60-61
周兴华. 2000. 矿渣中三氧化二硼的测定. 理化检验(化学分册), 36(10): 473

5.2　络合滴定设计实验

在络合滴定中，能否控制酸度进行滴定是首先要考虑的问题；其次，掩蔽剂的选择和应用是络合滴定成功的关键；指示剂的选择，要注意金属离子指示剂的酸碱性质和络合性质所造成的滴定误差。

实验 43 炉甘石洗剂中 ZnO 含量的测定

炉甘石洗剂是临床常用的外用药，其主要成分为氧化锌，还含有少量氧化铁、氧化镁、氧化钙、氧化锰等。炉甘石含量测定中多以氧化锌为标准，可以用络合滴定法对其中的氧化锌含量进行测定。

参考文献

吴晓平. 2016. 黄连汤制炉甘石的鉴别及含量测定方法. 中医临床研究，8(28): 31-32
杨慧，邵阳，陈昌云. 2013. 复方炉甘石洗剂中氧化锌含量测定方法探索. 南京晓庄学院学报，3: 59-62

实验 44 镍镁合金中镍、镁的测定

镍镁合金是由金属镍和金属镁通过高温熔炼得到的中间合金(二元合金)，一般镍的质量分数为 70%～90%，镁的质量分数为 10%～30%，同时含有少量的碳、硫、铁、铝、铜、锰等元素。目前，镍镁合金可用作镍氢电池负极材料；冶金上可作为球化剂用于生产轧辊的工作层和芯部，以改变石墨状态、增加轧辊的硬度。高含量镍、镁的测定方法主要采用 EDTA 滴定法。

参考文献

黎小阳，郭崇武，罗小平. 2018. 沉淀分离-滴定法测定碱性锌镍合金镀液中的锌含量. 电镀与涂饰, 37(1): 29-31
陆娜萍，年季强，张良芬，等. 2016. EDTA 滴定法测定镍镁合金中镍和镁. 冶金分析, 36(1): 62-66

实验 45 Mg^{2+}-EDTA 混合液中各组分的测定

首先，在 pH ≈ 10 溶液中，以 EBT 为指示剂，检查哪种组分过量。

(1)若 Mg^{2+}过量，移取一份试液用 EDTA 滴定过量的 Mg^{2+}。另取一份试液调 pH 至 5～6，用 XO 作指示剂，用 Zn^{2+}标准溶液滴定 EDTA 总量。

(2)若 EDTA 过量，移取一份试液调 pH 至 5～6，用 XO 作指示剂，用 Zn^{2+}标准溶液确定 EDTA 总量。另取一份试液，加 pH ≈ 10 NH_3-NH_4Cl 缓冲溶液，用 EBT 作指示剂，用 Zn^{2+}标准溶液滴定过量的 EDTA。

参考文献

武汉大学化学与分子科学学院实验中心. 2013. 分析化学实验. 2 版. 武汉: 武汉大学出版社, 101-102

5.3　氧化还原滴定设计实验

实验 46　合金中铬含量的测定

在硫磷混酸介质中，以硝酸银溶液作为催化剂，用过硫酸铵溶液将样品中的铬(III)氧化为铬(VI)。加入硫酸锰溶液，由于铬可先于锰被氧化，当 Mn^{2+} 被氧化至出现紫红色时，即可判断铬被氧化完全，锰同时被氧化为高锰酸。加入少量盐酸溶液并煮沸，破坏高锰酸。以苯基取代邻氨基苯甲酸溶液为指示剂，用硫酸亚铁铵标准溶液进行滴定。

测定铜合金中铬时，用硝酸溶解样品，冒高氯酸烟至锥形瓶瓶口一定时间，以苯基取代邻氨基苯甲酸溶液为指示剂，用硫酸亚铁铵标准溶液进行滴定。

参考文献

李传启，王露，杨崇秀. 2016. 高氯酸处理-硫酸亚铁铵滴定法测定铜合金中铬. 冶金分析, 36(4): 71-74
许洁瑜，麦丽碧，陈晓东. 2018. 微波消解-硫酸亚铁铵滴定法测定钴铬烤瓷合金中铬. 冶金分析, 38(4): 74-78

实验 47　加碘食盐中碘含量的测定

盐是人们日常生活中不可缺少的调味品。碘是人体新陈代谢和生长发育必需的微量营养元素，是人体合成甲状腺激素的主要原料。碘盐中的碘主要以 KIO_3 的形式存在，而 KIO_3 在酸性介质中能被 I^- 还原成 I_2，生成的 I_2 可以用硫代硫酸钠标准溶液进行滴定，以淀粉为指示剂，滴定至溶液的蓝色刚好消失为终点，从而求得碘盐中的碘含量。

参考文献

李松栋，张颖，张翠红，等. 2018. 加碘食盐中碘损失的研究. 山西化工, 38(2): 9-11

实验 48　含铁钢渣中氧化亚铁含量的测定

钢渣中铁的价态与相对分析对钢铁冶炼工艺的指导具有重要意义。钢渣中金属铁与亚铁的分离与测定通常是将金属铁转化为二价铁进入溶液后过滤分离。样品中的二价铁存在于不溶残留物中，经酸溶解后，采用重铬酸钾滴定法测定其中氧化亚铁的含量。

参考文献

贾香，邓慧兰，田晓照. 2017. 氧化还原滴定法测定含铁钢渣中氧化亚铁含量. 化学分析计量, 26(2): 89-91

5.4　沉淀滴定设计实验

实验 49　食品中氯化钠的测定

氯化钠存在于各种食品中，它对食品的保存时限及食品内部性状改变等都有重要的意义。GB/T 12457—2008《食品中氯化钠的测定》中规定了测定食品中氯化钠的方法，间接沉淀滴定法是其中一种。试液经酸化处理后，加入过量的硝酸银溶液，以硫酸铁铵为指示剂，用硫氰酸钾标准溶液滴定过量的硝酸银(福尔哈德法)。根据硫氰酸钾标准溶液的消耗量，计算食品中氯化钠的含量。也可以用直接沉淀滴定法，该法用硝酸银滴定氯离子，当氯离子完全被银离子沉淀后，过量的银离子与铬酸钾指示剂生成砖红色的铬酸银沉淀指示终点。根据硝酸银标准溶液的消耗量，计算食品中氯化钠的含量。

参考文献

严卓彦. 2016. 直接沉淀滴定法检测鱿鱼丝中氯化钠的不确定度评定. 食品安全质量检测学报, 7(7): 2785-2789
俞薇，韩丹，竺巧玲. 2015. 酱油中的氯化钠测定的能力验证结果与分析. 安徽农业科学，43(18): 306-307, 309
邹燕. 2010. 关于酱油中氯化钠的检测. 计量与测试技术, 1: 69-70

实验 50　复合肥料中氯离子含量的测定

复合肥料中氯离子的含量直接影响农作物的生长，尤其是忌氯的农作物，如甘薯、甜菜等对氯离子比较敏感，施肥时要严格控制用量。另外，长期使用氯离子含量高的复合肥料，土壤容易酸化和盐渍化，会影响种子发芽、出苗，抑制农作物生长。国家标准采用福尔哈德法测定复合肥料中的氯离子含量，即先加入过量的硝酸银标准溶液，使氯离子转化为氯化银沉淀，再用邻苯二甲酸二丁酯包裹沉淀后，以硫酸铁铵为指示剂，用硫氰酸铵标准溶液返滴定过量的硝酸银。

参考文献

杜颖，刘善江，陈益山. 2015. 有机肥料中氯离子检测方法的研究. 中国土壤与肥料, 1: 111-114
焦立为. 2006. 莫尔法沉淀滴定测定肥料中氯化物含量. 理化检验(化学分册), 42(3): 219-220
沈月，蔡玮. 2017. 复混肥料中氯离子含量测定的探讨. 浙江农业科学, 58(10): 1783-1784
周俊玲. 2005. 复混肥料中氯离子含量测量不确定度的评定. 化学分析计量, 14(5):7-10

实验 51　尼龙镀银纤维中银含量的测定

含银功能纤维作为一种新型纺织纤维，以其除臭、抗菌、抗静电、防辐射、医疗保健等功能得到广泛的关注和应用。含银功能纤维的功能性主要来源于所含的银，银含量是直接影响其功能性的关键指标，也是评定含银功能纤维质量、决定其价格的一项重要指标。测量镀银纤维中银含量可用沉淀滴定法，将镀银纤维用浓硫酸/浓硝酸高温溶解，使镀银纤维中的银变成硝酸银，基材纤维完全溶解。往得到的溶液中加入一定量的浓盐酸，反应生成氯化银白色沉淀。采用离心的方法将氯化银沉淀分离出来，再高温烘干，称出氯化银的质量，进而计算出镀银纤维样品中银含量。

参考文献

郭堃，马建伟，陈韶娟，等. 2015. 沉淀滴定法测定尼龙镀银纤维银含量. 山东纺织科技, 2: 27-29

5.5 分光光度法设计实验

实验 52 分光光度法测定钢样中的铬和锰

铬和锰都是钢中常见的有益元素，尤其在合金钢中应用比较广泛。铬和锰在钢中除以金属状态存在于固溶体中之外，还以碳化物(CrC_2、Cr_5C_2、Mn_3C)、硅化物(Cr_3Si、$MnSi$、$FeMnSi$)、氧化物(Cr_2O_3、MnO_2)、氮化物(CrN、Cr_2N)、硫化物(MnS)等形式存在。

试样经酸溶解之后，生成Mn^{2+}和Cr^{3+}，加入H_3PO_4掩蔽Fe^{3+}。在酸性条件下，以$AgNO_3$为催化剂，加入过量$(NH_4)_2S_2O_8$，将Cr^{3+}、Mn^{2+}分别氧化成$Cr_2O_7^{2-}$和MnO_4^-：

$$2Cr^{3+} + 3S_2O_8^{2-} + 7H_2O = Cr_2O_7^{2-} + 6SO_4^{2-} + 14H^+$$

$$2Mn^{2+} + 5S_2O_8^{2-} + 8H_2O = 2MnO_4^- + 10SO_4^{2-} + 16H^+$$

$Cr_2O_7^{2-}$在 420～450 nm 波长处吸收强烈，而MnO_4^-吸收很弱；MnO_4^-在 500～550 nm 波长处吸收强烈且产生双峰，而$Cr_2O_7^{2-}$吸收很弱。根据吸光度的加合原理，在$Cr_2O_7^{2-}$和MnO_4^-的最大吸收波长 440 nm 和 545 nm 处分别测定$Cr_2O_7^{2-}$和MnO_4^-混合溶液的吸光度，然后解联立方程，求出试液中铬和锰的含量。

参考文献

华中师范大学，东北师范大学，陕西师范大学，等. 2001. 分析化学实验. 3 版. 北京：高等教育出版社
武汉大学化学与分子科学学院实验中心. 2013. 分析化学实验. 2 版. 武汉：武汉大学出版社

实验 53 空气中微量氯气的测定

空气中氯气的主要来源是氯碱厂、氯的衍生物制备以及其他含氯化合物的合成等工业排出的废气。可用亚硫酸钠吸收空气中的氯气，氯气与亚硫酸钠反应生成氯离子，用过氧化氢将过量的亚硫酸钠氧化成硫酸根，在碱性条件下加热将过量的过氧化氢分解，然后利用硫氰酸汞法进行吸光度测定，最大吸收波长 460 nm。

参考文献

高小红，马玉龙，吴春姗，等. 2013. 吸光光度法测定空气中的微量氯气. 化学与生物工程，30(5)：85-87

实验 54　分光光度法测定植物叶中的铅含量

由于环境污染，植物叶(树叶、菜叶)上附有铅。铅是一种积累性毒物，过量铅对人体有很大危害。测定铅的常用方法为双硫腙显色、氰化钾掩蔽。该方法灵敏度高、选择性好，但引入剧毒物氰化钾又导致环境污染。也可用二甲酚橙显色，邻二氮菲为掩蔽剂，在 pH 为 4.5～5.4 条件下铅与二甲酚橙形成稳定的 1∶1 红色络合物，此络合物在 580 nm 波长处有最大吸收，摩尔吸光系数为 1.55×10^4 L/(mol·cm)。

参考文献

陈彰旭，方凤英，郑炳云，等. 2010. 植物叶上铅含量的吸光光度法测定. 临沂师范学院学报, 32(3): 105-109
杨光洁，郭洁，杜建华. 2000. 用二甲酚橙显色快速测定植物叶上的铅含量. 理化检验(化学分册), 36(9): 412-414

5.6　其他综合实验

实验 55　二草酸根合铜(Ⅱ)酸钾的制备及组成测定

实验原理

二草酸根合铜(Ⅱ)酸钾可以由硫酸铜与草酸钾直接混合制备，反应式如下：

$$2K_2C_2O_4 + CuSO_4 = K_2[Cu(C_2O_4)_2] + K_2SO_4$$

二草酸根合铜(Ⅱ)酸钾在水中的溶解度很小，但可以加入适量的氨水，使 Cu^{2+}形成铜氨离子而溶解(pH 约为 10)，也可用 NH_3-NH_4Cl 溶液溶解。

PAN 指示剂即 1-(2-吡啶偶氮)-2-萘酚，属于吡啶偶氮类显色剂，在 pH 为 1.9～12.2 呈黄色。PAN 指示剂与 Cu^{2+}的络合物为红色，而 EDTA 与 Cu^{2+}的络合物为蓝色。当滴定至终点时，游离态 PAN 自身的黄色与 EDTA-Cu 络合物的蓝色混合后形成翠绿色，即为终点的颜色。

实验用品

仪器：分析天平，水循环真空泵，烧杯(100 mL，400 mL，600 mL，800 mL)，量筒(10 mL，50 mL)，玻璃棒，酒精温度计，表面皿，电热板，布氏漏斗，抽滤瓶。

试剂：$CuSO_4 \cdot 5H_2O$(s)，$K_2C_2O_4 \cdot H_2O$(s)，$KMnO_4$ 标准溶液(0.02 mol/L)，EDTA 标准溶液(0.02 mol/L)，H_2SO_4 溶液(3 mol/L)，浓氨水，$Na_2C_2O_4$(s)，标准铜，NH_3-NH_4Cl 缓冲溶液，PAN 指示剂(0.1%，乙醇溶液)，HCl 溶液(6 mol/L)。

实验步骤

(1) 二草酸根合铜(Ⅱ)酸钾的制备。

称取 4 g $CuSO_4 \cdot 5H_2O$ 溶于 8 mL 90℃的水中，另称取 12 g $K_2C_2O_4 \cdot H_2O$ 溶于 44 mL 90℃的水中，趁热在剧烈搅拌下迅速将 $K_2C_2O_4$ 溶液加入 $CuSO_4$ 溶液中，冷至 10℃有沉淀析出，

减压过滤，用 8 mL 冷水分两次洗涤沉淀，50℃烘干产物。

(2) 铜含量的测定。

准确称取 0.17～0.19 g 产物，用 15 mL NH_3-NH_4Cl 缓冲溶液溶解，加 50 mL 蒸馏水。加 3 滴 PAN 指示剂，用 0.02 mol/L EDTA 标准溶液(详见实验 11)滴定至溶液由浅蓝色变为翠绿色，即为终点。

(3) 草酸根含量的测定。

准确称取 0.21～0.23 g 产物，用 2 mL 浓氨水溶解，加入 15 mL H_2SO_4 溶液(3 mol/L)，此时有淡蓝色沉淀出现。稀释至 100 mL，水浴加热至 70～85℃，趁热用 0.02 mol/L $KMnO_4$ 标准溶液(详见实验 20)滴定至微红色，1 min 内不褪色。沉淀在滴定过程中逐渐消失。

根据以上分析结果，计算产物中 Cu^{2+} 和 $C_2O_4^{2-}$ 的含量，并推算出产物的化学式。

参考文献

北京师范大学，东北师范大学，华中师范大学，等. 2014. 无机化学实验. 4 版. 北京：高等教育出版社

实验 56　硫代硫酸钠的制备及产物含量的测定

实验原理

$Na_2S_2O_3 \cdot 5H_2O$ 俗称大苏打，是无色透明的晶体，易溶于水，溶液显碱性，遇酸立即分解。硫代硫酸钠在分析化学中用来定量测定碘，在纺织工业和造纸工业中作脱氯剂，在摄影业中作定影剂，在医药行业中作急救解毒剂。

本实验采用 Na_2SO_3 和 S 在沸腾条件下直接化合法制备硫代硫酸钠：

$$Na_2SO_3 + S \xlongequal{\triangle} Na_2S_2O_3$$

室温下从溶液中结晶出来的硫代硫酸钠为 $Na_2S_2O_3 \cdot 5H_2O$。样品中硫代硫酸钠的含量可用 $K_2Cr_2O_7$ 作基准物质进行测定。$K_2Cr_2O_7$ 先与 KI 反应析出 I_2：

$$Cr_2O_7^{2-} + 6I^- + 14H^+ = 2Cr^{3+} + 3I_2 + 7H_2O$$

生成的 I_2 可用 $Na_2S_2O_3$ 溶液滴定，以淀粉溶液作指示剂，滴定至溶液的蓝色刚好消失即为终点。滴定反应为

$$I_2 + 2S_2O_3^{2-} = S_4O_6^{2-} + 2I^-$$

由消耗 $Na_2S_2O_3$ 溶液的体积和 $K_2Cr_2O_7$ 的质量可求出 $Na_2S_2O_3$ 溶液的浓度，再由称取 $Na_2S_2O_3$ 的质量求出 $Na_2S_2O_3$ 的含量。

实验用品

仪器：研钵，烧杯(100 mL)，漏斗，蒸发皿，水浴锅，分析天平，容量瓶(250 mL)，移液管(25 mL)，锥形瓶(750 mL)。

试剂：Na_2SO_3 (s)，$K_2Cr_2O_7$ (s)，$Na_2S_2O_3 \cdot 5H_2O$ (s)，硫粉，HCl 溶液(6 mol/L)，乙醇，KI 溶液(100 g/L)，淀粉溶液(5 g/L)。

实验步骤

(1) $Na_2S_2O_3 \cdot 5H_2O$ 的制备。

称取 2 g 硫粉，研碎后置于 100 mL 烧杯中，加入 1 mL 乙醇使其湿润。再加入 6 g Na_2SO_3 和 30 mL 去离子水，加热并搅拌至沸腾后改用小火加热，搅拌并保持微沸 40 min 以上，直至仅剩下少许硫粉悬浮在溶液中(此时溶液体积应不少于 20 mL，如太少，可在反应过程中适当补充水)。趁热过滤，将滤液转移至蒸发皿中，水浴加热，蒸发滤液直至溶液中有晶体析出时，冷却，即有大量晶体析出(如冷却时间较长而无晶体析出，可搅拌或加入一粒 $Na_2S_2O_3 \cdot 5H_2O$ 晶体以促使晶体析出)。减压过滤，并用少量乙醇洗涤晶体，抽干。40℃烘干(约 40 min)，称量，计算产率。

(2) $Na_2S_2O_3$ 标准溶液的配制。

准确称取 6～7 g $Na_2S_2O_3 \cdot 5H_2O$ 溶于少量刚煮沸并冷却的去离子水中，用刚煮沸并冷却的去离子水稀释至 250 mL。

(3) 产品中 $Na_2S_2O_3$ 含量的测定。

移取 25.00 mL $K_2Cr_2O_7$ 标准溶液(0.017 mol/L)于 250 mL 碘量瓶中，加入 5 mL KI 溶液(100 g/L)、5 mL HCl 溶液(6 mol/L)，加盖摇匀，在暗处放置 5 min。待反应完全，加入 100 mL 蒸馏水稀释，用 $Na_2S_2O_3$ 标准溶液滴定至呈浅黄绿色，加入 2～3 mL 淀粉溶液(5 g/L)，继续滴定至蓝色变为亮绿色即为终点，记录消耗 $Na_2S_2O_3$ 标准溶液的体积。平行测定三次。

根据记录的数据计算 $Na_2S_2O_3$ 标准溶液的浓度。由称取的 $Na_2S_2O_3 \cdot 5H_2O$ 的质量求出 $Na_2S_2O_3 \cdot 5H_2O$ 的百分含量。

参考文献

北京师范大学，东北师范大学，华中师范大学，等. 2014. 无机化学实验. 4 版. 北京：高等教育出版社
南京大学无机及分析化学实验编写组. 2015. 无机及分析化学实验. 3 版. 北京：高等教育出版社

实验 57 阻燃剂中氧化锌和三氧化二硼的测定

硼酸锌($2ZnO \cdot 3B_2O_3 \cdot 5H_2O$)是一种无机添加型阻燃产品，其质量标准通常以氧化锌和三氧化二硼的量表示。传统测定方法是将溶液 pH 控制在 7～10，铬黑 T 为指示剂，用 EDTA 滴定锌离子。再利用碳酸钠使锌离子沉淀分离，用硫酸使硼酸盐生成硼酸。以甲基红为指示剂，中和后加入甘露醇及酚酞指示剂，用标准碱溶液滴定，测定硼酸，进而计算出三氧化二硼的含量。也可以直接用络合滴定法测定锌，并求出氧化锌的含量，然后加入等摩尔络合滴定剂络合锌离子，然后测定硼酸，求出三氧化二硼的含量。

参考文献

任晓红. 2002. 阻燃剂硼酸锌的分析测定. 山西化工, 22(4)：33-34
王向东，龚晓光，段巧丽. 2004. 硼酸锌中三氧化二硼含量测定的新方法. 弹性体, 14(6)：61-63

附　录

附录 1　常用缓冲溶液的配制

缓冲溶液组成	pK	pH	配制方法
氨基乙酸-HCl	2.35 (pK_{a1})	2.3	取 150 g 氨基乙酸溶于 500 mL 水中后，加 80 mL 浓 HCl，稀释至 1 L
H_3PO_4-柠檬酸盐		2.5	取 113 g $Na_2HPO_4 \cdot 12H_2O$ 溶于 200 mL 水后，加 387 g 柠檬酸，溶解，过滤后稀释至 1 L
一氯乙酸-NaOH	2.86	2.8	取 200 g 一氯乙酸溶于 200 mL 水中，加 40 g NaOH，溶解后稀释至 1 L
邻苯二甲酸氢钾-HCl	2.95 (pK_{a1})	2.9	取 500 g 邻苯二甲酸氢钾溶于 500 mL 水中，加 80 mL 浓 HCl，稀释至 1 L
甲酸-NaOH	3.67	3.7	取 95 g 甲酸和 40 g NaOH 溶于 500 mL 水中，稀释至 1 L
NaAc-HAc	4.74	4.7	取 83 g 无水 NaAc 溶于水中，加 60 mL 冰醋酸，稀释至 1 L
六次甲基四胺-HCl	5.15	5.4	取 40 g 六次甲基四胺溶于 200 mL 水中，加 10 mL 浓 HCl，稀释至 1 L
Tris-HCl [三羟甲基氨基甲烷 $NH_2C(HOCH_2)_3$]	8.21	8.2	取 25 g Tris 试剂溶于水中，加 8 mL 浓 HCl，稀释至 1 L
NH_3-NH_4Cl	9.26	9.2	取 54 g NH_4Cl 溶于水中，加 63 mL 浓氨水，稀释至 1 L

附录 2　常用指示剂

1. 酸碱指示剂

指示剂	pH 变色范围	颜色变化	配制方法
甲基紫（第一变色范围）	0.13～0.5	黄～绿	1 g/L 或 0.5 g/L 水溶液
甲酚红（第一变色范围）	0.2～1.8	红～黄	0.04 g 指示剂溶于 100 mL 50%乙醇
甲基紫（第二变色范围）	1.0～1.5	绿～蓝	1 g/L 水溶液
百里酚蓝（第二变色范围）	1.2～2.8	红～黄	0.1 g 指示剂溶于 100 mL 20%乙醇
甲基紫（第三变色范围）	2.0～3.0	蓝～紫	1 g/L 水溶液
甲基橙	3.1～4.4	红～黄	1 g/L 水溶液
溴酚蓝	3.0～4.6	黄～蓝	0.1 g 指示剂溶于 100 mL 20%乙醇

续表

指示剂	pH 变色范围	颜色变化	配制方法
刚果红	3.0～5.2	蓝紫～红	1 g/L 水溶液
溴甲酚绿	3.8～5.4	黄～蓝	0.1 g 指示剂溶于 100 mL 60%乙醇
甲基红	4.4～6.2	红～黄	0.1 g 或 0.2 g 指示剂溶于 100 mL 60%乙醇
溴酚红	5.0～6.8	黄～红	0.1 g 或 0.04 g 指示剂溶于 100 mL 20%乙醇
溴百里酚蓝	6.0～7.6	黄～蓝	0.05 g 指示剂溶于 100 mL 20%乙醇
中性红	6.8～8.0	红～亮黄	0.1 g 指示剂溶于 100 mL 60%乙醇
酚红	6.8～8.0	黄～红	0.1 g 指示剂溶于 100 mL 20%乙醇
甲酚红	7.2～8.8	亮黄～茶红	0.1 g 指示剂溶于 100 mL 50%乙醇
百里酚蓝（第二变色范围）	8.0～9.6	黄～蓝	0.1 g 指示剂溶于 100 mL 20%乙醇
酚酞	8.2～10.0	无色～紫红	0.18 g 指示剂溶于 100 mL 60%乙醇
百里酚酞	9.3～10.5	无色～蓝	0.1 g 指示剂溶于 100 mL 90%乙醇

2. 酸碱混合指示剂

指示剂溶液的组成	变色点 pH	颜色		备注
		酸色	碱色	
三份 1 g/L 溴甲酚绿乙醇溶液 一份 2 g/L 甲基红乙醇溶液	5.1	紫红	绿	
一份 2 g/L 甲基红乙醇溶液 一份 1 g/L 甲基蓝乙醇溶液	5.4	红紫	绿	pH 5.2 红紫 pH 5.4 暗蓝 pH 5.6 绿
一份 1 g/L 溴甲酚绿钠盐水溶液 一份 1 g/L 氯酚红盐水溶液	6.1	黄绿	蓝紫	pH 5.4 蓝绿 pH 5.8 蓝 pH 6.2 蓝紫
一份 1 g/L 中性红乙醇溶液 一份 1 g/L 甲基蓝乙醇溶液	7.0	蓝紫	绿	pH 7.0 蓝紫
一份 1 g/L 溴百里酚蓝钠盐水溶液 一份 1 g/L 酚红钠盐水溶液	7.5	黄	绿	pH 7.2 暗绿 pH 7.4 淡紫 pH 6.2 深紫
一份 1 g/L 甲酚红钠盐水溶液 三份 1 g/L 百里酚蓝钠盐水溶液	8.3	黄	紫	pH 8.2 蓝绿 pH 8.4 蓝

3. 金属离子指示剂

指示剂	解离平衡和颜色变化	配制方法
铬黑 T (EBT)	$H_2In^- \xrightleftharpoons{pK_{a2}=6.3} HIn^{2-} \xrightleftharpoons{pK_{a3}=11.5} In^{3-}$ 紫红　蓝　橙	5 g/L 水溶液
二甲酚橙 (XO)	$H_3In^{4-} \xrightleftharpoons{pK=6.3} H_2In^{5-}$ 黄　红	2 g/L 水溶液
K-B 指示剂	$H_2In \xrightleftharpoons{pK_{a1}=8} HIn^- \xrightleftharpoons{pK_{a2}=13} In^{2-}$ 红　蓝　紫红 (酸性铬蓝 K)	0.2 g 酸性铬蓝 K 与 0.4 g 萘酚绿 B 溶于 100 mL 水中
钙指示剂	$H_2In^- \xrightleftharpoons{pK_{a2}=7.4} HIn^{2-} \xrightleftharpoons{pK_{a3}=13.5} In^{3-}$ 酒红　蓝　酒红	5 g/L 乙醇溶液
吡啶偶氮萘酚 (PAN)	$H_2In \xrightleftharpoons{pK_{a1}=1.9} HIn^- \xrightleftharpoons{pK_{a2}=12.2} In^{2-}$ 黄绿　黄　淡红	1 g/L 乙醇溶液
Cu-PAN (CuY-PAN 溶液)	$CuY+PAN+M^{n+} = MY+Cu\text{-}PAN$ 浅绿　无色　红色	在 10 mL 0.05 mol/L Cu^{2+}溶液中加 5 mL pH 5～6 的 HAc 缓冲溶液、1 滴 PAN 指示剂，加热至 60℃左右，用 EDTA 滴至绿色，得到约 0.025 mol/L CuY 溶液。使用时取 2～3 mL 于试液中，再加数滴 PAN 溶液
磺基水杨酸	$H_2In \xrightleftharpoons{pK_{a1}=2.7} HIn^- \xrightleftharpoons{pK_{a2}=13.1} In^{2-}$ 无色	10 g/L 水溶液
钙镁试剂	$H_2In^- \xrightleftharpoons{pK_{a2}=8.1} HIn^{2-} \xrightleftharpoons{pK_{a3}=12.4} In^{3-}$	5 g/L 水溶液

注：EBT、钙指示剂、K-B 指示剂等在水溶液中稳定性较差，可以配成指示剂与 NaCl 质量比为 1∶100 或 1∶200 的固体粉末。

4. 氧化还原指示剂

指示剂	$E^\ominus$ /V	颜色变化		配制方法
	$[H^+]$ = 1 mol/L	氧化态	还原态	
二苯胺	0.76	紫	无色	10 g/L 浓 H_2SO_4 溶液
二苯胺磺酸钠	0.85	紫红	无色	5 g/L 水溶液
N-邻苯氨基苯甲酸	1.08	紫红	红	0.1 g 指示剂加 20 mL 50 g/L Na_2CO_3 溶液，用水稀释至 100 mL
邻二氮菲-Fe(Ⅱ)	1.06	浅蓝	红	1.485 g 邻二氮菲加 0.965 g $FeSO_4$ 溶解，稀释至 100 mL (0.025 mol/L 水溶液)
5-硝基邻二氮菲-Fe(Ⅱ)	1.25	浅蓝	紫红	1.608 g 5-硝基邻二氮菲-Fe(Ⅱ)加 0.695 g $FeSO_4$ 溶解，稀释至 100 mL(0.025 mol/L 水溶液)

5. 吸附指示剂

指示剂	配制	用于测定		
		可测元素(括号内为滴定剂)	颜色变化	测定条件
荧光黄	1%钠盐水溶液	Cl^-，Br^-，I^-，SCN^-(Ag^+)	黄绿～粉红	中性或弱碱性
二氯荧光黄	1%钠盐水溶液	Cl^-，Br^-，I^-(Ag^+)	黄绿～粉红	pH = 4.4～7.2
四溴荧光黄(曙红)	1%钠盐水溶液	Br^-，I^-(Ag^+)	橙红～红紫	pH = 1～2

附录 3　常用酸碱溶液的浓度和密度

试剂	密度/(g/mL)	w/%	c/(mol/L)
盐酸	1.18～1.19	36～38	11.6～12.4
硝酸	1.39～1.40	65.0～68.0	14.4～15.2
硫酸	1.83～1.84	95～98	17.8～18.4
磷酸	1.69	85	14.6
高氯酸	1.68	70.0～72.0	11.7～12.0
冰醋酸	1.05	99.8(优级纯)，99.5(分析纯)，99.0(化学纯)	17.4
氢氟酸	1.13	40	22.5
氢溴酸	1.49	47.0	8.6
氨水	0.88	25.0～28.0	13.3～14.8

附录 4　常用基准物质及其干燥条件与应用

基准物质		干燥后组成	干燥条件，t/℃	标定对象
名称	分子式			
碳酸氢钠	$NaHCO_3$	Na_2CO_3	270～300	酸
碳酸钠	$Na_2CO_3 \cdot 10H_2O$	Na_2CO_3	270～300	酸
硼砂	$Na_2B_4O_7 \cdot 10H_2O$	$Na_2B_4O_7 \cdot 10H_2O$	放在含 NaCl 和蔗糖饱和溶液的干燥器中	酸
碳酸氢钾	$KHCO_3$	K_2CO_3	270～300	酸
草酸	$H_2C_2O_4 \cdot 2H_2O$	$H_2C_2O_4 \cdot 2H_2O$	室温空气干燥	碱或 $KMnO_4$
邻苯二甲酸氢钾	$KHC_8H_4O_4$	$KHC_8H_4O_4$	110～120	碱
重铬酸钾	$K_2Cr_2O_7$	$K_2Cr_2O_7$	140～150	还原剂
溴酸钾	$KBrO_3$	$KBrO_3$	130	还原剂
碘酸钾	KIO_3	KIO_3	130	还原剂
铜	Cu	Cu	室温干燥器中保存	还原剂

续表

基准物质		干燥后组成	干燥条件，t/℃	标定对象
名称	分子式			
三氧化二砷	As_2O_3	As_2O_3	室温干燥器中保存	氧化剂
草酸钠	$Na_2C_2O_4$	$Na_2C_2O_4$	130	氧化剂
碳酸钙	$CaCO_3$	$CaCO_3$	110	EDTA
锌	Zn	Zn	室温干燥器中保存	EDTA
氧化锌	ZnO	ZnO	900～1000	EDTA
氯化钠	NaCl	NaCl	500～600	$AgNO_3$
氯化钾	KCl	KCl	500～600	$AgNO_3$
硝酸银	$AgNO_3$	$AgNO_3$	280～290	氯化物
氨基磺酸	$HOSO_2NH_2$	$HOSO_2NH_2$	在真空 H_2SO_4 干燥器中保存 48 h	碱
氟化钠	NaF	NaF	铂坩埚中 500～550℃下保存 40～50 min 后，H_2SO_4 干燥器中冷却	

附录 5　常用化合物的分子量

分子式	分子量	分子式	分子量	分子式	分子量
Ag_3AsO_4	462.52	CO_2	44.01	$CuCl_2 \cdot 2H_2O$	170.48
AgBr	187.77	$CO(NH_2)_2$	60.06	CuI	190.45
AgCl	143.32	CaO	56.08	$Cu(NO_3)_2$	187.56
AgCN	133.89	$CaCO_3$	100.09	$Cu(NO_3)_2 \cdot 3H_2O$	241.60
Ag_2CrO_4	331.73	CaC_2O_4	128.10	CuO	79.545
AgI	234.77	$CaCl_2$	110.99	Cu_2O	143.09
$AgNO_3$	169.87	$CaCl_2 \cdot 6H_2O$	219.08	CuS	95.61
AgSCN	165.95	$Ca(NO_3)_2 \cdot 4H_2O$	236.15	$CuSO_4$	159.60
$AlCl_3$	133.34	$Ca(OH)_2$	74.09	$CuSO_4 \cdot 5H_2O$	249.68
$AlCl_3 \cdot 6H_2O$	241.43	$Ca_3(PO_4)_2$	310.18	CuSCN	121.62
$Al(NO_3)_3$	213.00	$CaSO_4$	136.14	$FeCl_2$	126.75
$Al(NO_3)_3 \cdot 9H_2O$	375.13	$CdCO_3$	172.42	$FeCl_2 \cdot 4H_2O$	198.81
Al_2O_3	101.96	$CdCl_2$	183.32	$FeCl_3$	162.21
$Al(OH)_3$	78.00	CdS	144.47	$FeCl_3 \cdot 6H_2O$	270.30
$Al_2(SO_4)_3$	342.14	$Ce(SO_4)_2$	332.24	$FeNH_4(SO_4)_2 \cdot 12H_2O$	482.18
$Al_2(SO_4)_3 \cdot 18H_2O$	666.41	$Ce(SO_4)_2 \cdot 4H_2O$	404.30	$Fe(NO_3)_3$	241.86
As_2O_3	197.84	$CoCl_2$	129.84	$Fe(NO_3)_3 \cdot 9H_2O$	404.00
As_2O_5	229.84	$CoCl_2 \cdot 6H_2O$	237.93	FeO	71.846
As_2S_3	246.02	$Co(NO_3)_2$	182.94	Fe_2O_3	159.69
$BaCO_3$	197.34	$Co(NO_3)_2 \cdot 6H_2O$	291.03	Fe_3O_4	231.54
BaC_2O_4	225.35	CoS	90.99	$Fe(OH)_3$	106.87
$BaCl_2$	208.24	$CoSO_4$	154.99	FeS	87.91
$BaCl_2 \cdot 2H_2O$	244.27	$CoSO_4 \cdot 7H_2O$	281.10	Fe_2S_3	207.87
$BaCrO_4$	253.32	$CrCl_3$	158.35	$FeSO_4$	151.90
BaO	153.33	$CrCl_3 \cdot 6H_2O$	266.45	$FeSO_4 \cdot 7H_2O$	278.01
$Ba(OH)_2$	171.34	$Cr(NO_3)_3$	238.01	$FeSO_4 \cdot (NH_4)_2SO_4 \cdot 6H_2O$	392.13
$BaSO_4$	233.39	Cr_2O_3	151.99	H_3AsO_3	125.94
BiCl	315.34	CuCl	98.999	H_3AsO_4	141.94
BiOCl	260.43	$CuCl_2$	134.45	H_3BO_3	61.83

续表

分子式	分子量	分子式	分子量	分子式	分子量
HBr	80.912	K_2SO_4	174.25	$Na_2S \cdot 9H_2O$	240.18
HCN	27.026	$MgCO_3$	84.314	$NaSCN$	81.07
$HCOOH$	46.026	MgC_2O_4	112.33	Na_2SO_3	126.04
CH_3COOH	60.052	$MgCl_2$	95.211	Na_2SO_4	142.04
H_2CO_3	62.025	$MgCl_2 \cdot 6H_2O$	203.30	$Na_2S_2O_3$	158.10
$H_2C_2O_4$	90.035	$Mg(NO_3)_2 \cdot 6H_2O$	256.41	$Na_2S_2O_3 \cdot 5H_2O$	248.17
$H_2C_2O_4 \cdot 2H_2O$	126.07	$MgNH_4PO_4$	137.32	$NiCl_2 \cdot 6H_2O$	237.69
HCl	36.461	MgO	40.304	NiO	74.69
HF	20.006	$Mg(OH)_2$	58.32	$Ni(NO_3)_2 \cdot 6H_2O$	290.79
HI	127.91	$Mg_2P_2O_7$	222.55	NiS	90.75
HIO_3	175.91	$MgSO_4 \cdot 7H_2O$	246.47	$NiSO_4 \cdot 7H_2O$	280.85
HNO_3	63.013	$MnCO_3$	114.95	P_2O_5	141.94
HNO_2	47.013	$MnCl_2 \cdot 4H_2O$	197.91	$PbCO_3$	267.20
H_2O	18.015	$Mn(NO_3)_2 \cdot 6H_2O$	287.04	PbC_2O_4	295.22
H_2O_2	34.015	MnO	70.937	$PbCl_2$	278.10
H_3PO_4	97.995	MnO_2	86.937	$PbCrO_4$	323.20
H_2S	34.08	MnS	87.00	$Pb(CH_3COO)_2$	325.30
H_2SO_3	82.07	$MnSO_4$	151.00	$Pb(CH_3COO)_2 \cdot 3H_2O$	379.30
H_2SO_4	98.07	$MnSO_4 \cdot 4H_2O$	223.06	PbI_2	461.00
$Hg(CN)_2$	252.63	NH_3	17.03	$Pb(NO_3)_2$	331.20
$HgCl_2$	271.50	CH_3COONH_4	77.083	PbO	223.20
Hg_2Cl_2	472.09	NH_4Cl	53.491	PbO_2	239.20
HgI_2	454.40	$(NH_4)_2CO_3$	96.086	$Pb_3(PO_4)_2$	811.54
$Hg_2(NO_3)_2$	525.19	$(NH_4)_2C_2O_4$	124.10	PbS	239.30
$Hg_2(NO_3)_2 \cdot 2H_2O$	561.22	$(NH_4)_2C_2O_4 \cdot H_2O$	142.11	$PbSO_4$	303.30
$Hg(NO_3)_2$	324.60	NH_4HCO_3	196.01	SO_2	64.06
HgO	216.59	$(NH_4)_2HPO_4$	132.06	SO_3	80.06
HgS	232.65	$(NH_4)_2MoO_4$	79.055	$SbCl_3$	228.11
$HgSO_4$	296.65	NH_4NO_3	80.043	$SbCl_5$	299.02
Hg_2SO_4	497.24	$(NH_4)_2S$	68.14	Sb_2O_3	291.50
$KAl(SO_4)_2 \cdot 12H_2O$	474.38	NH_4SCN	76.12	Sb_2S_3	339.68
KBr	119.00	$(NH_4)_2SO_4$	132.13	SiF_4	104.08
$KBrO_3$	167.00	NH_4VO_3	116.98	SiO_2	60.084
KCl	74.551	NO	30.006	$SnCl_2$	189.62
$KClO_3$	122.55	NO_2	46.006	$SnCl_2 \cdot 2H_2O$	225.65
$KClO_4$	138.55	Na_3AsO_3	191.89	$SnCl_4$	260.52
KCN	65.116	$Na_2B_4O_7$	201.22	$SnCl_4 \cdot 5H_2O$	350.596
K_2CO_3	138.21	$Na_2B_4O_7 \cdot 10H_2O$	381.37	SnO_2	150.71
K_2CrO_4	194.19	$NaBiO_3$	279.97	SnS	150.776
$K_2Cr_2O_7$	294.18	$NaCN$	49.007	$SrCO_3$	147.63
$K_3Fe(CN)_6$	329.25	Na_2CO_3	105.99	SrC_2O_4	175.64
$K_4Fe(CN)_6$	368.35	$Na_2CO_3 \cdot 10H_2O$	286.14	$SrCrO_4$	203.61
$KFe(SO_4)_2 \cdot 12H_2O$	503.24	$Na_2C_2O_4$	134.00	$Sr(NO_3)_2$	211.63
$KHC_2O_4 \cdot H_2O$	146.14	CH_3COONa	82.034	$Sr(NO_3)_2 \cdot 4H_2O$	283.69
$KHC_2O_4 \cdot H_2C_2O_4 \cdot 2H_2O$	254.19	$CH_3COONa \cdot 3H_2O$	136.08	$SrSO_4$	183.68
$KHC_4H_4O_6$	188.18	$NaCl$	58.443	$UO_2(CH_3COO)_2 \cdot 2H_2O$	424.15
$KHSO_4$	136.16	$NaClO$	74.442	$ZnCO_3$	125.39
KI	166.00	$NaHCO_3$	84.007	ZnC_2O_4	153.40
KIO_3	214.00	$Na_2HPO_4 \cdot 12H_2O$	358.14	$ZnCl_2$	136.29
$KIO_3 \cdot HIO_3$	389.91	$Na_2H_2Y \cdot 2H_2O$	372.24	$Zn(CH_3COO)_2$	183.47
$KMnO_4$	158.03	$NaNO_2$	68.995	$Zn(CH_3COO)_2 \cdot 2H_2O$	219.50
$KNaC_4H_4O_6 \cdot 4H_2O$	282.22	$NaNO_3$	84.995	$Zn(NO_3)_2$	189.39
KNO_3	101.10	Na_2O	61.979	$Zn(NO_3)_2 \cdot 6H_2O$	297.48
KNO_2	85.104	Na_2O_2	77.978	ZnO	81.38
K_2O	94.196	$NaOH$	39.997	ZnS	97.44
KOH	56.106	Na_3PO_4	163.94	$ZnSO_4$	161.44
$KSCN$	97.18	Na_2S	78.04	$ZnSO_4 \cdot 7H_2O$	287.54

主要参考文献

北京师范大学, 东北师范大学, 华中师范大学, 等. 2014. 无机化学实验. 4 版. 北京: 高等教育出版社

大连理工大学分析化学教研室. 2008. 分析化学(双语版). 大连: 大连理工大学出版社

华中师范大学, 东北师范大学, 陕西师范大学, 等. 2011. 分析化学(上册). 4 版. 北京: 高等教育出版社

华中师范大学, 东北师范大学, 陕西师范大学, 等. 2015. 分析化学实验. 4 版. 北京: 高等教育出版社

黄应平. 2012. 分析化学实验(英汉双语教材). 武汉: 华中师范大学出版社

王新宏. 2009. 分析化学实验(双语版). 北京: 科学出版社

魏明霞, 马明广. 2013. 分析化学(双语版). 北京: 化学工业出版社

武汉大学. 2011. 分析化学实验(上册). 5 版. 北京: 高等教育出版社

武汉大学. 2016. 分析化学(上册). 6 版. 北京: 高等教育出版社

武汉大学化学与分子科学学院实验中心. 2013. 分析化学实验. 2 版. 武汉: 武汉大学出版社

Hage D S, Carr J D. 2012. 分析化学和定量分析(英文版). 北京: 机械工业出版社

Ham B M, MaHam A. 2015. Analytical Chemistry: A Chemist and Laboratory Technician's Toolkit. New York: Wiley

Iwunze M O. 2005. Laboratory Experiments in Analytical Chemistry. AuthorHouse

Kenkel J. 2014. Analytical Chemistry for Technicians. 4th ed. Boston: CRC Press

Ready A V R, Swain K K, Venkatesh K. 2012. Experiments in Analytical Chemistry. India: Association of Environmental Analytical Chemistry of India

Skoog D A, West D M, Holler F J, et al. 2004. Fundamentals of Analytical Chemistry. London: Thomson Brooks/Cole